'Mathematics is an important part of most branches of engineering. Students who ignore mathematics at high school level may lose several job opportunities in the future because mathematics is used in every area and has many career options. Strong curricula that incorporates various applications of mathematics into engineering is lacking in most African universities. This book has tried to address various gaps and provide various strategies that inspire our young generation.'

Dr Getachew Temesgen, Tshwane University of Technology, South Africa

'This book on mathematical modelling and engineering design, which comes out after the knowledge exchange from the third AfricaLics Conference in Oran (Algeria), and other AfricaLics academies, is very inspiring. Having been engaged in the process, I know that the book will be very useful to stimulate quality research – the results of which will continue to address the challenges in the African context and to fill a gap on this very important topic. I strongly recommend that the book is integrated into the curriculum of the African education system. It is a knowledge- and resource-rich book that is timely and essential to promote African knowledge.'

Professor Abdelkader Djeflat, Professor of Innovation Studies at Lille University, France and Chair, The Maghtech Network (www.maghtech.org)

'I highly recommend this book for scholars and researchers involved in science, technology, engineering and mathematics (STEM) education and research. Mathematical modelling and engineering design together offer a valuable analytical tool to engage STEM-related challenges. The diversity of the contributions shows the richness scholars and researchers will be able to draw on. This is an easy-to-read book, providing multiple perspectives for the reader. In a post-COVID-19 context, a book as this can offer both interdisciplinary and multidisciplinary perspectives to better engage problems. The book will surely enrich a global audience.'

Dr Emmanuel Ojo, University of the Witwatersrand, Johannesburg, South Africa

Engineering Design and Mathematical Modelling

Engineering Design and Mathematical Modelling: Concepts and Applications consists of chapters that span the engineering design and mathematical modelling domains.

Engineering design and mathematical modelling are key tools/techniques in the science, technology and innovation spheres. Whilst engineering design is concerned with the creation of functional innovative products and processes, mathematical modelling seeks to utilize mathematical principles and concepts to describe and control real world phenomena. Both can be useful tools for spurring and hastening progress in developing countries. They are also areas where Africa needs to 'skill-up' in order to build a technological base.

The chapters in this book cover the relevant research trends in the fields of both engineering design and mathematical modelling. This book was originally published as a special issue of the *African Journal of Science, Technology, Innovation and Development.*

Nnamdi Nwulu is a researcher, educationist and engineer. He holds BSc and MSc degrees in Electrical and Electronic Engineering and a PhD degree in Electrical Engineering. He is currently Associate Professor in the Department of Electrical and Electronic Engineering Science at the University of Johannesburg, South Africa. He is also a Professional Engineer registered with the Engineering Council of South Africa (ECSA), a National Research Foundation (NRF) rated researcher, a Senior Member of the Institute of Electrical and Electronics Engineers (IEEE), a Senior Research Associate in the SARChI Chair in Innovation Studies at the Tshwane University of Science and Technology, South Africa and Associate Editor of the *African Journal of Science, Technology, Innovation and Development* (AJSTID).

Mammo Muchie founded the *African Journal on Science, Technology, Innovation and Development* (AJSTID) in 2008 and has served as its Editor-in-Chief ever since. He has been given the Best Institutional Senior Researcher of the Year Merit and the Academic Excellence Award. Professor Muchie's scholarly contribution to the discipline of innovation has been in strengthening and contextualising the theoretical framework of National Innovation Systems as applied to the African context. Without his work, the principles and theory of the National Innovation Systems, as they are practised in the developed and industrial economies, could not be applied directly to the developing or under-developed and largely agrarian economies of the African continent.

Engineering Design and Mathematical Modelling

Concepts and Applications

Edited by

Nnamdi Nwulu and Mammo Muchie

Routledge
Taylor & Francis Group

LONDON AND NEW YORK

First published 2021
by Routledge
2 Park Square, Milton Park, Abingdon, Oxon OX14 4RN

and by Routledge
52 Vanderbilt Avenue, New York, NY 10017

Routledge is an imprint of the Taylor & Francis Group, an informa business

British Library Cataloguing in Publication Data
A catalogue record for this book is available from the British Library

ISBN 13: 978-0-367-47732-5

Typeset in Times
by Newgen Publishing UK

Publisher's Note
The publisher accepts responsibility for any inconsistencies that may have arisen
during the conversion of this book from journal articles to book chapters, namely
the inclusion of journal terminology.

Disclaimer
Every effort has been made to contact copyright holders for their permission to
reprint material in this book. The publishers would be grateful to hear from any
copyright holder who is not here acknowledged and will undertake to rectify any
errors or omissions in future editions of this book.

Contents

Citation Information

The following chapters in this book were originally published in the *African Journal of Science, Technology, Innovation and Development*, volume 11, issue 3 (May 2019). When citing this material, please use the original page numbering for each article, as follows:

For any permission-related enquiries please visit:
www.tandfonline.com/page/help/permissions

Notes on Contributors

Akintunde Samson Alayande, Department of Electrical and Electronic Engineering Science, University of Johannesburg, South Africa and Department of Electrical and Electronics Engineering, University of Lagos, Nigeria.

Clinton O. Aigbavboa, Sustainable Human Settlement and Construction Research Centre, Faculty of Engineering and the Built Environment, University of Johannesburg, South Africa.

Obiora Cornelius Collins, Institute of Systems Science, Durban University of Technology, South Africa.

Uyikumhe Damisa, Department of Electrical and Electronic Engineering Science, University of Johannesburg, South Africa.

Hailemichael Teshome Demissie, School of Law, University of Gondar, Ethiopia.

Kevin Jan Duffy, Institute of Systems Science, Durban University of Technology, South Africa.

Solomon O. Giwa, Department of Mechanical Engineering, Olabisi Onabanjo University, Ago Iwoye, Nigeria and Department of Mechanical and Aeronautical Engineering, Faculty of Engineering, University of Pretoria, South Africa.

Njabulo Kambule, Department of Geography, Environmental Management and Energy Studies, University of Johannesburg, South Africa.

Lawrence Joseph Kerefu, Department of Mechanical Engineering, College of Engineering and Technology, Kibamba, Dar es Salaam, Tanzania.

Abayomi T. Layeni, Department of Mechanical Engineering, Olabisi Onabanjo University, Ago Iwoye, Nigeria.

Juliana Zawadi Machuve, Department of Mechanical and Industrial Engineering, College of Engineering and Technology, Kibamba, Dar es Salaam, Tanzania.

Nyasha Mahonye, School of Economic and Business Sciences, University of Witwatersrand, Johannesburg, South Africa.

Golden Makaka, Department of Physics, University of Fort Hare, Alice, South Africa.

Sampson Mamphweli, Department of Physics, University of Fort Hare, Alice, South Africa.

Peace-maker Masukume, Department of Physics, University of Fort Hare, Alice, South Africa.

Charles Mbohwa, School of Mechanical and Industrial Engineering, University of Johannesburg, South Africa.

Mammo Muchie, DST/NRF SARChI Research Professor on Science, Technology and Innovation Studies, Tshwane University of Technology, Pretoria, South Africa.

Musole Innocent Muheme, Department of Electrical and Electronic Engineering Science, University of Johannesburg, South Africa.

Patrick Mukumba, Department of Physics, University of Fort Hare, Alice, South Africa.

Kabeya Musasa, Department of Electrical and Electronic Engineering Science, University of Johannesburg, South Africa.

Ibrahim Niankara, College of Business Administration, Al Ain University of Science and Technology, Abu Dhabi, United Arab Emirates.

Collins N. Nwaokocha, Department of Mechanical Engineering, Olabisi Onabanjo University, Ago Iwoye, Nigeria and Department of Mechanical and Aeronautical Engineering, Faculty of Engineering, University of Pretoria, South Africa.

Nnamdi Nwulu, Department of Electrical and Electronic Engineering Science, University of Johannesburg, South Africa.

Olusegun A. Oguntona, Sustainable Human Settlement and Construction Research Centre, Faculty of Engineering and the Built Environment, University of Johannesburg, South Africa.

Thokozani Silas Simelane, Institute of Systems Science, Durban University of Technology, South Africa.

Yanxia Sun, Department of Electrical and Electronic Engineering Science, University of Johannesburg, South Africa.

Kowiyou Yessoufou, Department of Geography, Environmental Management and Energy Studies, University of Johannesburg, South Africa.

Tatenda Zengeni, Competition and Tariff Commission of Zimbabwe, Harare, Zimbabwe.

Acknowledgements

We would like to thank the DST/NRF SARChI (Science, Technology, Innovation and Development) at the Tshwane University of Technology, Pretoria, South Africa for all the hard work to produce the special issue and the book project.

We recognise and appreciate the support given by the Taylor & Francis team, in particular Caroline Church and Eleanore Reinders. Thanks also to the AfricaLics network for the grant to support the engineering design research and AfricaLics for the grant to undertake the engineering design research that the SARChI chair received.

We also would like to thank the *African Journal of Science, Technology, Innovation and Development* (AJSTID) and extend our fullest appreciation and thanks to all the contributing authors.

Dedication

The African contribution of the knowledge heritage to the world is still not recognised. We dedicate this book to all the African ancestors who left a rich knowledge and science legacy with their original contributions to mathematics, astronomy, physics and all the sciences. We dedicate this in particular to the founder and originator of the great philosopher Zera Yacob who created the African Enlightenment long before the European Enlightenment.

Preface

Exposure to the African Roots of Mathematics Modelling and Engineering Design

Mammo Muchie and Nnamdi Nwulu

Africa is often seen as the world's science, mathematics and engineering Sahara desert. This book is much needed to open up the research on mathematics and engineering, focusing on modelling and design through a historical examination of the much-ignored science, mathematics and engineering that originated in Africa. The claim that the mathematics and science originating in Africa is pre-scientific must be challenged. Mathematical modelling and engineering design must challenge the epistemological ethnocentrism that degrades Africa as the mathematics and engineering dry zone. If Africa is the origin of humanity and its negative data started with slavery over 500 years ago, surely all the thousands of years of history cannot be empty of knowledge? We need to look back to feed forward from that long memory any of the learning we can excavate from mathematics, physics, chemistry, astronomy and engineering. Some of our findings demonstrate that there is a rich African knowledge heritage.

Mathematics modelling in Africa has been mixed across the arts, riddles, architecture and games. All of the following have originated in ancient Africa: mathematical games, artefacts, digital computer systems, facial and pattern recognition, **geometrical thinking, geometry and symmetry, mathematical understanding,** design sense, construction skills, mathematical recognition of food sources in the bush, and the African Stonehenge calendar.

Some examples of mathematics in ancient Africa include the following: Lebombo bone (South Africa Swaziland) ~35,000 BC; Ishango bone (Congo) ~25,000 BC; Ancient Egypt: Ahmose (Rhind) mathematical papyrus (π). Sona geometry (Angola) – which is symmetrical and monolinear – has been the basis for new mathematical ideas such as mirror curves and various classes of matrices (cycle, cylinder, helix) (Gerdes, 2009). African fractals (West and Central Africa) involve repeating geometric patterns at various scales and are used in textiles, paintings, sculptures, cosmologies, architecture and town planning. Fractals are the basis of the worldwide web (Eglash and Odumosu, 2005). Taint taboos affect the **numerical expression of numbers**. For example, the number "7" is tied to the origins of creation and great divinities within the Niger Delta region. Accordingly, to express the number "7", you would say "6+1". The Yoruba also created their own **complex counting system** based on units of 20 (instead of 10)

In ancient mathematics in Ethiopia, Africans used the binary logic of mathematical calculation, which is similar to today's internet browsers and other computer-based systems (see www.youtube.com/watch?v=OOKp9_sSkZg).

Recently, a very revealing and remarkable method of mathematical calculation was revealed by BBC Four with contributions by commentators from the Open University in the UK (see www.google.co.za/search?q=ancient+ethiopian+mathematics&biw=1366&hih=673&thm=isch&tbo=u frequres uni.Quu—](Q.i -]1 K'Vi/p/ 1]AV/ABPfoULAg&sqi—2&.vc il—IK'HFQsAQordpr—1). There are now two global economies: real and virtual. The binary logic of mathematics used for the internet world was used in Ethiopian mathematics.

Example 1: 7 × 8 = 56
    14 × 4 = 56
    28 × 2 = 56
    56 × 1 = 56

Example 2: 9 × 16 = 144
    18 × 8 = 144
    36 × 4 = 144
    72 × 2 = 144
    144 × 1 = 144

Africans also have originated many scientific discoveries. Evidence has been acknowledged by astronomers and engineers from Western research universities that for many major inventions, documented contributions have been made by Africans. Carbon steel was made 2000 years ago in Tanzania; astronomical observations in Mali by the Dogons have been acknowledged by Carl Segall of Cornell University. Also, in each of the following fields, Africans have been inventors: language, mathematical systems, architecture, agriculture, cattle-rearing, navigation of inland waterways and open seas, medicine, communication, and writing systems.

African knowledge in agriculture is rich. This was recognised and fully acknowledged by the US Academy of Science in 1996:

> Africa has more native cereals than any other continent. It has its own species of rice, as well as finger millet, fonio, peral millet, sorghum, teff, guinea millet, and several dozen wild cereals whose grains are eaten. This is a food heritage that has fed people for generations, possibly stretching back to the origins of mankind. It is also a local legacy of genetic wealth upon which a sound food future might be built. But strangely, it has been bypassed in modern times... Forward thinking scientists are starting to look at the old cereal heritage with unbiased eyes. Peering past the myths, they see waiting in the shadows a storehouse of resources whose qualities offer promise not just to Africa, but the world. The Ethiopian crop teff is medicine. It has still great potential to be the nutrition for healthy living. Food security in Africa should not be in doubt at all.
>
> (Muchie et al, 2003, pp 43–61)

Finally, the African origin of science should not be ignored. It can be acknowledged as the foundation of Western science. Civilisation, astronomy, science, mathematics and philosophy originated from Africa. There is a real challenge for Africans to understand that the difficulties of today and tomorrow can be transformed into beauty after tomorrow if they are prepared to look back to the deep and rich cultural contributions made by their ancestors. Mathematical modelling and engineering design from African roots need to be on the research agenda, with more discoveries waiting to be accomplished.

References

Bernal, Martin. 1985. *Black Athena: The Afroasiatic Roots of Classical Civilisation*. Rutgers University Press Classics.

Chirikure, Shadreck, Rob Burrett, and Robert B. Heimann. 2009. "Beyond Furnaces and Slags: A Review Study of Bellows and their Role in Indigenous African Metallurgical Processes." *Azania: Archaeological Research in Africa* 44(2):195–215.

Diop, Cheikh Anta. 1974. *The African Origin of Civilization: Myth or Reality*. Lawrence Hill Books.

Diop, Cheikh Anta. 1987a. *Black Africa: The Economic and Cultural Basis for a Federated State*. Lawrence Hill Books. Original edition: *Les fondements économiques et culturels d'un état fédéral d'Afrique noire*.

Diop, Cheikh Anta. 1987b. *Precolonial Black Africa*. Lawrence Hill Books.

Diop, Cheikh Anta. 1991. *Civilization of Barbarism: An Authentic Anthropology*. Lawrence Hill Books.

Eglash, Ron, and Toluwalogo B. Odumosu. 2005. "Fractals, Complexity and Connectivity in Africa." In *What Mathematics from Africa?*, edited by Giandomenica Sico. Polimetrica Publisher.

Fabayo, J.A. 1996. "Technological Dependence in Africa: Its Nature, Causes, Consequences and Policy Derivatives." *Technovation* 16(7):357–370.

Mind Research Institute. n.d. "13 Interesting Facts About Math in Ancient Africa." https://blog.mindresearch.org/blog/math-facts-ancient-africa

Muchie, Mammo. 2009. "Some Questions Related to the Epistemological Purity of Science" In *Encyclopedia of Life Support Systems*, edited by Prasada Reddy. EOLSS Publishers. UNESCO. www.eolss.net/sample-chapters/c15/e1-31-04-01.pdf

Muchie, Mammo, Peter Gammeltoft, and Bengt-Åke Lundvall. 2003. *Putting Africa First: The Making of African Innovation Systems*. Aalborg University Press, pp. 43–61.

Wolpert, Lewis. 1998. *The Unnatural Nature of Science*. Harvard University Press.

INTRODUCTION

Engineering design and mathematical modelling: Concepts and applications

Engineering design and mathematical modelling are key tools/techniques in the Science, Technology and Innovation spheres. Whilst engineering design is concerned with the creation of functional innovative products and processes, mathematical modelling seeks to utilize mathematical principles and concepts to describe and control real world phenomena. Both of these can be useful tools for spurring and hastening development in developing countries. They are also areas where Africa needs to 'skill-up' in order to build her technological base. This special issue of the African Journal of Science, Technology, Innovation and Development (AJSTID) contains 13 original articles that cover relevant research trends in the fields of both engineering design and mathematical modelling. These articles can be broadly classified into status of engineering design in African nations (3 papers) and applications of mathematical modelling in the following areas: emissions modelling (1 paper), education (1 paper), urban city dynamics (2 papers), finance (1 paper) and energy (5 papers).

The article by Hailemichael Teshome Demissie, 'Current state and trajectory of Design Engineering in Kenya' explores the state of engineering design in Africa and highlights the present challenges in the country. The author highlights the shortage of engineers in the country and the use of archaic engineering design tools as key challenges that plague the engineering design landscape in Kenya. Lawrence Joseph Kerefu and Juliana Zawadi Machuve in their article 'Students' perception of engineering design for competitiveness in Africa: The case of Tanzania, East Africa' investigate the perception of students regarding the quality of the engineering education they receive in the country's tertiary institutions and the proportion that explicitly handles engineering design. Empirical evidence obtained points to the need for a review of the curricula in line with the severe engineering challenges of the country. In 'Barriers hindering biomimicry adoption and application in the construction industry', Olusegun Oguntona and Clinton Aigbavboa discuss biomimicry which is an engineering design philosophy that studies nature or biological processes and mimics them in the design of engineering systems and products. The approach has found wide application and has the potential to enable sustainability in many industries, but has yet to find widespread deployment in the African context. The authors used the South African construction industry as a case study and argued that four factors or barriers hamper the widespread deployment of biomimicry in the construction industry, namely: information and technology-related barriers, risk and cost-related barriers, knowledge-related barriers and regulation-related barriers. It can also be argued that these factors affect the widespread deployment of engineering design techniques on the African continent.

There are numerous papers in this special issue that deal with applications of mathematical modelling approaches that can be used to understand, predict and control various phenomena of interest. Solomon O. Giwa, Collins N. Nwaokocha and Abayomi T. Layeni in 'Inventory of kiln stacks emissions and health risk assessment: Case of a cement industry in Southwest Nigeria' utilized mathematical modelling techniques to estimate the quantity of emissions from a cement factory in southwest Nigeria. A comparison between the air quality index of the cement plant and the World Health Organization (WHO) standards was also performed. Their results show that the amount of emission from the cement industry is more than the recommended limit and constitutes a health hazard.

In his paper, Ibrahim Niankara focused on the education sector in 'Modelling the effects of exposure to risk on junior faculty productivity incentives under the academic tenure system' and deployed concepts from probability theory and economics of risk to model outputs under the academic tenure system. The major premise is that scientific outputs can be modelled like any other form of production; consequently, faculty scientific output can be quantified. However, as publications are requirements for academic staff promotion, it has a corresponding effect on the risks faculty face as publications hold some inbuilt uncertainty. Increasing values of this uncertainty has an effect of reducing faculty research incentives. The author is of the opinion, in view of modelling results, that tenure track rules for junior faculty members should be well thought out and implemented in order to simultaneously maintain faculty productivity and institutional reputation.

Cities are a key component of developing nations and have a significant effect on national and regional economic development. Mathematical modelling tools can be used to understand underlying dynamics of cities which is necessary for effective planning and to mitigate urban poverty. The two papers by O. C. Collins, T. S. Simelane and K. J. Duffy study population dynamics and socioeconomic dynamics in key urban cities on the African continent. Their first paper 'Analyses of mathematical models for city population dynamics under heterogeneity' studies the population dynamics in three African cities and investigates the impact of heterogeneity in income and expenditure on cities residents. The results show that income and expenditure can have different but significant effects on the population dynamics of different population groups (students, workers, visitors, business people, and job seekers). In their second paper, 'Mathematical model showing how socioeconomic dynamics in African cities could widen or reduce inequality' the authors investigate socioeconomic dynamics in African cities considering income, employment and educational opportunities and their effect on social inequality. They arrive at the conclusion that policymakers should actively pursue synergism between various socioeconomic classes in order to stem the tide of worsening inequality in African cities.

Mathematical modelling definitely plays an important role in the financial sector. Nyasha Mahonye and Tatenda Zengeni in 'Exchange rate impact on output and inflation: A historical perspective from Zimbabwe' investigated the effect of exchange rate fluctuations on real output growth and inflation in Zimbabwe. Using historical data and regression analysis, the author determines that exchange rate fluctuations have an impact in the short and long term on real output growth. It was also discovered that exchange rate fluctuations have no impact on inflation in the short run but have an impact in the long run. Without a doubt, these results have significant policy implications especially for export policy.

Recent trends in the energy sphere have seen increased penetration of renewable energy resources (RES) in the electrical grid. They have also seen the advent of prosumers (electricity end-users who simultaneously produce and consume electricity). Uyikumhe Damisa, Nnamdi Nwulu and Yanxia Sun in 'A mathematical formulation of the joint economic and emission dispatch problem of a renewable energy-assisted prosumer microgrid' determine how to optimize operations in a setup involving prosumers with renewables whilst minimizing cost and emissions. The authors opine that incentives have a ripple effect on minimizing emissions in a RES prosumer setup. Another recent trend that has seen increased penetration in modern energy systems is the rise of electric vehicles (EVs). EVs have however found limited penetration in the African continent in spite of their environmental benefits. Kabeya Musasa, Musole Muheme, Nnamdi Nwulu and Mammo Muchie in their article 'A simplified control scheme for electric vehicle-power grid circuit with DC distribution and battery storage systems' consider a pure battery EV with a direct current circuit and battery storage system (BSS). The developed control scheme yields an acceptable performance in spite of parameter variations in the EV-circuits and the impact of the BSS charging/discharging process. Observing the DC voltage characteristics indicates no overshoot or ringing whilst the steady-state error is kept to zero.

Njabulo Kambule, Kowiyou Yessoufou, Nnamdi Nwulu and Charles Mbohwa in 'Temporal analysis of electricity consumption for prepaid metered low- and high-income households in Soweto, South Africa' utilized historical data and regression analysis to perform a temporal analysis of prepaid metered electricity consumption in Soweto, South Africa. Both low-income and high-income consumers were analyzed and the key research finding was that electricity consumption had decreased by 48% since the inception of prepaid meters; however, 60% of household income is spent on electricity bills. This has important policy implications as 60% of income spent on electricity connotes energy poverty. The authors call for special measures to protect energy poor households.

In 'A novel approach for the identification of critical nodes and transmission lines for mitigating voltage instability in power networks', Akintunde Alayande and Nnamdi Nwulu develop a novel method for determining critical power system components (nodes and lines) in order to maintain a power system stability. Their approach is based on the graph theory approach and has practical applications, especially on the African continent where there are frequent occurrences of blackouts and brown outs.

'Design, construction and mathematical modelling of the performance of a biogas digester for a family in the Eastern Cape province, South Africa' by Patrick Mukumba, Golden Makaka, Samposn Mamphweli and Peacemaker Masukume mathematically modelled the expected yield from a $1m^3$ biogas digester with donkey dung as the substrate. The biogas digester was also constructed, and results obtained from the constructed prototype show a strong agreement with the mathematical model. The designed biogas digester should find many practical uses, especially on the African continent.

Viewed holistically, the papers in this volume seek to unearth important insights about various facets of life in developing countries. Utilizing tools from engineering design and mathematical modelling, they proffer policy recommendations useful in the economic, healthcare and financial sectors. Taken together, they can help lead to an improvement in the lives of people in the developing nations of the world.

Acknowledgments

We extend our gratitude to the Swedish International Development Agency (SIDA), the African Network for Economics of Learning, Innovation, and Competence Building Systems (AfricaLics), AJSTID's Chief Editor, Anga Baskaran, and the editorial office for their support towards this special issue. We also thank the reviewers for volunteering their time and submitting their reviews timeously.

ORCID

Mammo Muchie http://orcid.org/0000-0003-4831-3113

Nnamdi Nwulu

Mammo Muchie

Current state and trajectory of design engineering in Kenya

Hailemichael Teshome Demissie

In the last few decades, the methods and models of engineering design practices have changed drastically due to the introduction of better technology and sophisticated design tools. Engineers are now moving away from manual design and drafting using A4 paper sheets, rulers and tri-squares to the use of computer-aided engineering tools in designing and manufacturing. These new advances in engineering design practices have led to huge increases in efficiency gains in engineering practices that, in turn, result in better infrastructure and, ultimately, economic growth and social development. Kenya needs to pay more attention to the changes in technology and expertise in the field of engineering to improve the nation's economy. Even though Kenya, and Africa at large, are growing rapidly in technology and in the field of engineering design, there is still a big chasm between the continent and the developed countries in the practice and education of engineering design. This paper argues that more should be done in accelerating the uptake of engineering design technologies in Kenya and by the same token in the wider Africa.

Introduction

Design is such an indispensable element of engineering that engineering itself is defined in terms of design: engineering without design is simply a contradiction in terms. While some simply equate design to engineering (Elliott 2010, 54) others place emphasis on both by referring to 'design and engineering' albeit using both terms alternatively and inseparably (Bell 2007). The media often prefer design to engineering to refer to the same notion (UNESCO 2010, 63).

The congruence of the ideas represented by the two terms, engineering and engineering design (ED), becomes even more seamless when their respective definitions are juxtaposed for comparison. A widely quoted definition of ED is provided by the US Accreditation Board for Engineering and Technology (ABET):

> Engineering design is the process of devising a system, component, or process to meet desired needs. It is a decision making process (often iterative), in which the basic sciences, mathematics, and engineering sciences are applied to convert resources optimally to meet a stated objective (cited in Eide et al. 2002, 79).

Engineering design has been recognized as the essence of engineering and a core component of the innovation capacity building process. It is a crucial driver of economic development in any nation – developed or developing alike. The design process is the core activity in any engineering field and product development. It is the most critical stage of the value creation process in any product development. It is also the stage where 80% of the environmental impact of physical or service products including infrastructure development is determined, according to Britain's Design Council (Antonelli 2008, 23). Most definitions of engineering, it is argued, downplay the role of design and obscure its integral role in engineering. While engineering involves all the activities listed in any definition of engineering like making, building, operating, and sustaining, it is design that dictates the direction these activities take. (Turnbull 2010, 30)

Kenyan law recognizes the centrality of engineering design in engineering in general. *The Engineering Act of 2011* defines engineering as

> the creative application of scientific principles to design or develop structures, machines, apparatus, or manufacturing processes, or works utilizing them singly or in combination or to construct or operate the same with full cognizance of their design or to forecast their behavior under specific operating conditions or aspects of intended functions, economics of operation and safety to life and property;

A more recent piece of legislation, *The Engineering Technologists and Technicians Bill* 2015 simply retains the definition in *The Engineering Act of 2011*, defining 'engineering technology' as

> the use of scientific knowledge to solve engineering problems using the creative application of scientific principles to design, construct and develop structures, machines, equipment or manufacturing processes or works utilizing them singly or in combination or to construct or operate the same with full cognizance of their design or to forecast their behavior under specific operating conditions or aspects of intended functions, economics of operation and safety to life and property

These definitions essentially describe engineering in terms of design. Any discussion on engineering education and practice in any country involves a substantial discussion on engineering design in particular. This is an important statement because the state of engineering design can explain the state of engineering in general. This paper seeks to achieve this – it seeks to explain the state of engineering in Kenya through the prism of the state of engineering design in Kenya and vice versa.

Methodology

The methodology employed for the present paper is mainly exploratory. The paper is therefore dedicated to a description and analysis of the state of engineering design in Kenya and the dynamics of its current and future development. The research work that resulted in the present paper drew on publicly available publications, government documents, secondary and grey literature and internet-based desktop work. Unstructured interviews with a few selected engineering school deans and lecturers, students, and architectural and construction firms were conducted to get an overview of the state of engineering design in Kenya and what the future holds for the development of engineering design practice in Kenya and the wider region. Direct observations from visits to the engineering schools and the firms provided valuable first-hand information supporting some of the findings of the paper. The information was retrieved in real-time and it involved people and organizations that are currently operating and are directly involved on the ground in this field of research.

The role of engineering and engineering design in Kenya's national development plan

Engineering is recognized as one of the most important focus areas that Kenya has identified as a pillar of its national development plan. The country's national development plan known as *Vision 2030* was launched in 2008 and lays out the roadmap to propel the country to industrial development, and acquire competence in manufacturing, infrastructure development and overall economic growth. Kenya is aiming at attaining the status of 'a globally competitive and prosperous nation with a high quality of life by 2030' (*Vision 2030*). To implement the long-term plan in *Vision 2030*, two Medium Term Plans were launched. The first Medium Term Plan (MTP I) was a five year plan for the 2008–2012 period while the second (MTP II) is for the period from 2012 to 2017.

Among the major infrastructure projects under *Vision 2030* are the $350 m Thika Highway that was completed in 2013, the $200 m Nairobi Southern Bypass, the $635 m Jomo Kenyatta International Airport redevelopment project, and the $2.66bn Mombasa-Nairobi 480 km standard gauge railway. The construction of Lamu port and the expansion of the port at Mombasa, the ambitious Konza City project – a project to build a Technology City that is to become the IT business hub, 'Africa's Silicon Savannah', are also *Vision 2030* initiatives. The Lamu port project deserves a special mention here as part of the LAPPSET project, the Lamu Port South Sudan Ethiopia Transport Corridor project. The project is for the construction of a $23bn port and oil refinery at Lamu and the transport infrastructure extending to Ethiopia and South Sudan that includes an oil pipeline, a railway line and a motorway. The LAPSSET project is one of the major projects that elucidates how significant the role of the engineering profession is in Kenya's national development plan – a plan that aims further, beyond the borders of Kenya.

The hugely ambitious infrastructure component of *Vision* 2030 is revealing the deficit in engineering capability in the country. To realize the infrastructure development of *Vision 2030*, Kenya needs 30,000 more engineers. The current state of engineering education and training leaves much to be desired and the future is full of challenges. However, there are opportunities that can be used to overcome the possible challenges.

The state of engineering design in Kenya

Kenya is still making laws not merely to regulate the engineering profession but also to rid it of its colonial legacy and bring it to the heights of modern, global engineering standards. This is indicative of the fact that the engineering profession is merely emergent in Kenya. The legacy of engineering practices and codes from the colonial era are still persistent. The engineering education and the institutions are reminiscent of the colonial times. The building codes, until the changes introduced in 1995, were retained from the British colonial system and have irrelevant requirements such as the roof of a house being able to carry a certain snow load in a country that has never seen any snowfall (Erastus and Wuchuan 2014). The codes also had design requirements that made it impossible to use local materials. To address such discrepancies between colonial engineering practices and institutions, Kenya is enacting laws and adopting policies that are responsive to the current level of global engineering practices and technology.

The objectives of these laws, as stated in them, is to raise the standards of engineering practice. These laws draw on Kenya's experience under previous laws. *The Engineering Act of 2012* repeals and replaces the *Engineers Registration Act Cap 350*. The new law establishes the Engineering Board of Kenya that has the responsibility to regulate the engineering professional services, to oversee the development and general practice of engineering, to register engineers and firms, and to set standards. Among its functions are the approval and accreditation of engineering courses in public and private universities and other tertiary level educational institutions offering education in engineering. The Board is also entrusted with the establishment of a school of engineering and the provision of facilities and opportunities for learning, professional exposure and skills acquisition, and holding continuing professional development programmes for engineers. The law also envisages the establishment of the Kenya Academy of Engineering and Technology whose purpose shall be to advise the National and the County Governments on policy matters relating to engineering and technology.

Since it started to operationalize its functions, the Board has ruled on the accreditation of engineering programmes in Kenyan universities – an exercise that resulted in the denial of accreditation of most programmes in the country. This is a clear sign that the engineering capability in Kenya is at a very low level. The accreditation row has revealed a gap between the country's needs for engineering capacity and its ability to meet the need. The denial of accreditation by the Board was a serious blow to the country's ambitious plan to plug the skills gap by raising the number of engineers to implement *Vision 2030*. Kenya needs 30,000 more engineers to implement the vision of becoming a middle-income country by 2030.[1]

As of November 2014, there were 8700 trained engineers in Kenya with only 2000 licensed to practise (Muindi 2014). The tables below show the number of engineers according to EBK and to membership of IEK (Tables 1 and 2).

Table 1: Engineers Registration Board register as at May 2012.

Category	Gender		Total
	Male	Female	
Registered consulting engineers	272 (98.2%)	5 (1.8%)	277
Registered engineers	1298 (96.8%)	43 (3.2%)	1341
Reg. graduate engineers	4974 (92.3%)	413 (7.7%)	5387
Graduate technicians	1128 (98.5%)	17 (1.5%)	1145

To raise the number of engineers to the level required for the implementation of *Vision 2030*, Kenya needs nothing short of a revolution in its engineering education system. With most of the engineering programmes in nearly all universities declared inadequate, Kenya is unlikely to plug the shortfall of 30,000 engineers before 2030. This is further compounded by the low levels of enrolment in STEM education and the migration of the experienced engineers to greener pastures looking for better opportunities. The suspension of courses in almost all universities by the EBK exercising its powers under *The Engineering Act of 2011* does not paint a promising picture of the future of engineering education in Kenya

The EBK declined the requests for the approval of courses offered by the universities citing that they do not meet the required standards. Out of the 22 public universities, only four public universities – University of Nairobi, Moi University, Jomo Kenyatta University and Egerton University – were given approval to offer some courses. The University of Nairobi, the oldest and largest university in Kenya, was given the green light to offer only five courses: Civil, Agricultural, Mechanical, Electrical and Electronics engineering. On the other hand, Kenyatta University, the second largest university, was denied accreditation for all its engineering courses. However, Kenyatta University has somehow managed to get accreditation for some of its courses after a protracted renegotiation with the Board.

While the decision by the Board not to approve most of the courses was welcomed by some, including the Institute of Engineers of Kenya, the learned society of the engineering profession in Kenya, it is not a decision that has gained universal assent. Some engineers are sceptical of the Board's decision as can be seen from the comment below:

> The board has become a hindrance to training more engineers in Kenya. They are not allowing in new courses outside those they were themselves taught in, and it seems they fear flooding the market with new modern-age engineers. (Kennedy Kimathi, a drilling engineer with a geothermal power company in Nairobi, quoted in Waruru 2015)

Following the decision by the Engineers Board of Kenya denying accreditation of engineering courses, the universities are taking steps to fulfil the standards set by the Board. The universities are hiring lecturers from around the world, and they are acquiring lab equipment and machinery. They have yet to stimulate enrolment in engineering – an activity that has to start early with STEM education in early schooling.

The row over accreditation of university courses is an expression of many underlying developments in Kenya. Notable among these developments are the rapid growth of universities, their autonomy in decisions on their programmes and the rise of a rather assertive Engineers Board of Kenya with its statutory mandate to regulate the engineering profession that gave rise to a dual accreditation system: the Board's powers are in tension with the autonomy of universities exercised by the universities and the powers of the government body charged with the responsibility of accrediting university programmes – the Commission for University Education (CUE) (Muindi 2014). These developments showcase the dynamics of the evolution of the engineering profession in Kenya that is still in flux.

Engineering design – evolving out of the old way

The practice of engineering design has changed dramatically in the last three decades. The introduction of computers to the engineering design process, with such applications as CAD, CAM and AutoCAD, is recognized to be of no less significance than the steam engine and the electric motor (Edquist and Jacobson 1988) The design engineer today cannot be imagined without his/her computer. It was not long ago that the designer had to go around with a messy set of tools including a clumsy drawing board, T-squares, triangles, rulers and compasses, rolls of drawing paper, pencils and pens. While this is still the case in many places in Africa, the computer has replaced most, if not all, of these instruments with added functionalities, ease of use and increased productivity. Computer-Aided-Design (CAD) is the norm today and it is simply a tool that designers cannot afford not to have and continue as designers. Over the decades, CAD software has been further refined and developed providing 2D and 3D drawings, calculations on materials and bills. The advantages of CAD and its variants are too many to list here in a work of such brevity.

With CAD, the designer gets a virtual image of the product to be developed and this is easily communicated to other members of the design team around the world as the design process is now a collaborative effort in most cases. CAD has made it easier to provide the design to other participants in the product development process, particularly the manufacturers and marketing professionals who may want to modify the product design. With the help of CAD, any proposed modifications can be done on the screen almost instantly without the need to go to the drawing board again and start from scratch.

While CAD with its variants is now commonplace among engineering practitioners, it cannot be said that it has been fully utilized in African engineering practice. It is probably striking for many to read about the need to raise the computer literacy of engineers in Africa. It is a reasonable expectation that engineers should have more

Table 2: The Institute of Engineers of Kenya records as at May 2012.

Category	Gender		Total
	Male	Female	
Fellows	81 (98.8%)	1 (1.2%)	82
Corporate engineers	1465 (98.6%)	21 (1.4%)	1486
Associates	99 (100%)	0	99
Graduates	1569 (96.6%)	55 (3.4%)	1624

than average proficiency in the use of ICT tools including database searches and the internet. In some cases, this expectation is not met as this is not even considered in existing engineering curricula (Onwuka 2009, 336). When and where computer skills should be taught is also an issue when comparing the US and Nigeria. In Nigeria it is observed that considerable time is spent on teaching computer skills to engineering students while no such teaching occurs in the US, since engineering students have already covered it in high school (Falade 2006). In Ghana, there are challenges that can be generalized to many African countries as well: inadequate infrastructure and support – the intermittent electricity supply being the major problem – low levels of computer literacy and use, lack of hardware, low ratio of networked computers, high local internet rates, and low band-width capability of internet services (Appiagyei and Addai-Mensah 2006). These challenges are not unique to the teaching and practice of engineering only, and are to be addressed under the broad topic of the digital divide. Yet, it has to be underscored that the engineering profession has a bigger stake in the amelioration of the digital divide.

While CAD is rightly referred to as the 'vehicle of design', this should not obscure the significance of the driver of the vehicle (Jordaan 2006). The importance of the training of the engineering designer is even more pronounced with the use of tools like CAD. Computer-aided-design does not dispense with the designer's role at least for the near and mid-term future. The designer is afforded more power, time and other resources with the use of CAD but cannot retire to the backroom leaving the job for the computer. The job remains the designer's albeit with the help of CAD software. This brings the designer into an even more glaring spotlight.

In Kenya, universities and colleges offering engineering courses still teach students the traditional means of using pen, pencil, paper, rulers and other measuring and layout tools to create engineering designs, because the complex computer-aided technology is not easily acquired by the schools. They are considered expensive equipment. The number of required tools and the amount of equipment owned by schools is insufficient for the hundreds of engineering students (and thousands in some universities). Only major universities and well-known colleges are able to acquire this equipment. In addition, this technology has only recently (less than 10 years ago) been introduced to the updated engineering syllabus in universities and colleges due to the lack of technology access, the poor economic state of the country as well as the low level of expertise of university lecturers. Although universities are introducing more machines and tools, students are more likely to get better training and skills during

their attachments to or internships in engineering firms, architecture companies, and construction and energy firms. During their internship and attachments, students get full and adequate exposure to engineering design technology, more time learning about it, and better qualified trainers to teach them.

The main reason universities and colleges still teach students the traditional means to create engineering designs, using pen, pencil, paper, rulers and other measuring and layout tools, is because these tools are easily available to the students and more affordable.[2] Out in the field, engineering companies have fully embraced the use of computer-aided design technology because it is more efficient, faster and more accurate. The current taxation of engineering design software applications and equipment is now favourable; hence, more companies are taking this up. Mr. Samuel Gitau, the Director of Planning Systems Services Limited, agrees that use of this technology to manufacture products of great quality and in large quantities makes it economically effective (personal communication). This is because the design and product analysis are more accurate; hence, less time and material are used to make prototypes and final products. As a result, companies manufacture more products and do this in short lead times.

For a long while, computers were too costly in Kenya. Only well-established and wealthy companies and individuals could afford them. However, with time, computers have become a lot more affordable in the region. Computer-aided design and drafting systems have therefore started gaining popularity. Currently, this design technology is locally available and can be purchased without much difficulty. Of late, more companies in Kenya are more receptive to engineering design technology and tools. Computers and software applications are readily available, thus allowing engineers and other professionals the use of these graphic capabilities. In addition, improved design technology is included in the country's school syllabus, and, as a result, the potential to generate a higher number of people making up a more skilled labour force is expanding.

Professionals have turned to computer-aided design graphic software systems due to their favourable characteristics: they are menu driven (have commands for selection), the data is interactive, it is easy to manipulate, to add or delete, it displays both graphic and textual material on the screens, and it can produce software to hardcopy (Asby and Jones 1994). Basically, the key features of a CAD system comprises 2-dimensional and 3-dimensional drafting and modelling tools, the storage and retrieval of parts and materials, and the provision for engineering calculations, the engineering design and layout. Another feature of the computer-aided design system is the circuit and logic analysis and simulation, which enable the act of easily modifying the layout and design and operation of the system (Derbyshire 2010).

The most commonly used computer-aided design tool in Kenya is AutoCAD. This is because it is more affordable and easily available to Kenyan firms. In addition, compared to other software, AutoCAD is user friendly and the interface is easier to comprehend and manipulate. It has prominent features such as the drawing area, space used to draw design, and the ribbon, where the user selects tools to draw with (Omura and Benton 2012, 9). The common variants of this technology include AutoCAD LT, AutoCAD 360 (formerly AutoCAD WS) and the student version. AutoCAD can be used for drawing designs, plotting, and creating orthographic views from its solid model (Pohit and Ghosh 2007, 8).

AutoCAD is a widely used software application for both 2D and 3D computer-aided design and drafting. A bigger percentage of the AutoCAD software applications used in Kenya is of the 2D kind. However, more engineering companies have started to embrace the 3D technology in the country. The AutoCAD technology is used by engineers, project managers, architects, graphic designers, and related professionals. This technology is easily available and can be purchased online. Other computer-aided and drafting tools and technology used in Kenya include ArchiCAD, 3D CAD/CAM tools, Atir, 3D printers, Bentley software, Computational Fluid Dynamics (CFD) for design, Matlab and Revit software.

Challenges of engineering design in Kenya
Engineering design in Kenya is emergent and still evolving, trying to keep pace with the ever-evolving technology. Even though Kenya has been rather slow in embracing the technology and digitization of engineering design practices, the country has made great progress in the last two years. The engineering curriculum of universities and colleges has been overhauled, and schools are getting better teachers, buying more engineering machines, and moving away from the traditional means of designing. This is also reflected in the fieldwork and engineering firms in Kenya, where computer-aided design and drafting technology is being widely utilized. However, not all universities and colleges have upgraded to the sophisticated design machines, as these are mainly available to the major universities.

Engineering education is currently facing many challenges including poor funding for the acquisition of software and hardware, inadequate facilities, and inadequate human capacity. The main challenge in the field is skills development. Kenya has a number of universities and colleges that offer engineering courses, which facilitate the learning of engineering design subjects and give students exposure to the tools and technology. However, a great deal of capacity building work remains for the universities and colleges as evidenced by the recent accreditation row. The universities have taken steps to meet the required standards in engineering education by hiring qualified and capable lecturers from around the world.

While capacity building at the tertiary education level is certainly important, tackling the lack of interest in STEM education in early education is critical. The low level of enrolment in STEM education will derail the plans for raising the number of engineers for the implementation of the Kenya's *Vision 2030*. The severe gender imbalance in the engineering profession in general shown in the tables above and in STEM education enrolment needs to be addressed if Kenya is to remain on track with its development plan.

The challenge of human capital development in the field of engineering is getting even more complicated as more and more projects are carried out by foreign firms, especially Chinese firms that the government is attracting with its 'Go East' policy (Oxford Business Group 2014). Foreign firms have secured nearly half of the market in current infrastructure projects and have participated in projects ranging from the Thika Highway to Konza City. The engineering, architecture and construction firms are among the most important learning sites for engineers, especially the young. Kenya is likely to miss out on the opportunity to develop local engineering design capability due to the prominent roles foreign firms play in the infrastructure projects of the country. Kenya used to have a regulation that required foreign firms to have 30% local ownership. That regulation was scrapped in 2012 and foreign firms can take projects on more favourable terms to them compared to the situation under the previous requirement of including local professionals and firms. Foreign firms are now required to sub-contract 30% of work and materials to local firms by the new regulator – National Construction Authority (NCA) (Oxford Business Group 2014).

With foreign companies occupying such a large space in the infrastructure projects in the country, the room for learning for local firms becomes narrower. The country, therefore, should put more effort into training and developing local expertise so as to support local engineering firms. Local engineering design capability in particular needs to be actively supported, as it is critical for the entire engineering profession. Educating and training the now generation of engineers benefits the future of the country (Roessner 2008).

Africa's latecomer advantage in engineering design practice and Kenya's opportunities
Obviously, Africa is not at the forefront of developing or widely using most of the engineering design technologies. Africa will be disadvantaged if it tries to shun the 'technologisation' of engineering design on the basis of excuses of affordability or prioritizing other exigencies. Technologies have made engineering design a much easier activity and Africa aspiring to build capacity in the field has the opportunity to exploit these technologies to its advantage. CAD and other design technologies are technologies that are simple to use. These technologies are not designed for Europe or any other specific region. Hence, there is no wisdom in reinventing the wheel trying to go through what others have arduously gone through. Furthermore, these technologies have become indispensable to the learning and practice of engineering design to the extent

that engineering design without computers or CAD software simply does not make sense.

The challenge in Africa is not that of the inadaptability of these technologies to Africa's condition but the lack of supporting infrastructure like the supply of electricity and lack of access to web services and electronic hardware. Above all, it is also a challenge of getting concept champions who will promote and popularize the technologies and their benefits. Lobbying governments to procure these technologies and to negotiate access to these technologies from suppliers on favourable terms is among the tasks that must be carried out successfully if Africa is to benefit from these technologies. For this to be realized, policy changes are required.

An overarching policy stance that Africa should consider is how to approach engineering design capacity building. As Martin Bell (2007) highlighted:

> A particularly important component of that change must centre on what are described here as 'design and engineering' capabilities. These need to be distinctly identified in their own right in policy analysis and debate – not simply treated as adjuncts to, or qualified sub-species of, Research and Development.

Bell explains why 'design and engineering', in his preferred terminology, has to be placed at the centre of policy by elaborating the pivotal role engineering design plays and the linkages that are tethered to it. On the one hand, engineering design is an activity that translates knowledge created through R&D into practical use and socio-economic benefits, and on the other, it links production and markets 'back' to the creation of new knowledge in R&D. Understanding this pivotal role of engineering design for policymakers in Africa is key in devising growth strategies. A lack of understanding of engineering issues among policymakers in Africa emerged as a major finding of a study by the Royal Academy of Engineers (2012). The study did not even consider the 'technologisation' of engineering design as its focal point and was generally addressing the engineering sector. As such, the Royal Academy study did not highlight the emphasis Bell (2007) put on 'design and engineering'. Owing to the very low base of engineering in Africa, the studies by Bell and the Royal Academy did not find it necessary to single out engineering design as emphatically as this paper seeks to argue.

In trying to achieve a better understanding of engineering issues among policymakers in Africa, it is argued here that the integral role of engineering design should be given priority. Africa's disadvantage is huge due to the poor level of engineering education and practice, and manufacturing capabilities that made the continent dependent on other economies. Africa has been outsourcing both the design and manufacturing components of engineering jobs to other economies for several decades. It cannot survive let alone compete in the global economy without recovering this long-lost ground by starting to 'insource' these jobs. Developing its own engineering design capacity is a choice it cannot avoid. Engineering solutions brought from elsewhere are not always relevant; nor are they in the best interest of Africa. As some commentators observed, 'Africa is littered with wells and pumps that do not work' (UNESCO 2010). Such foreign engineering solutions have cost the continent dearly both in financial terms and in terms of the negative impact on the environment. Africa needs to develop its own indigenous engineering design capacity and this cannot be addressed without reorienting engineering design education towards a vigorous programme of capacity building that takes into account the lightning speed at which the 'technologisation' of engineering design, design theories, instrumentation and disciplines are changing.

The technologies of engineering design hold the promise of stimulating engineering education in Kenya and other African countries. A major problem in encouraging students to enrol in STEM education and engineering is the fact that they consider these fields abstract and difficult to relate to their daily lives. The comment by a University of Nairobi professor aptly summarizes this:

> Technological gaps will keep on growing between developed and developing countries as long as new approaches are not adopted in the teaching of sciences and mathematics in primary and secondary schools. Most Kenyan secondary school laboratories are ill-equipped for students to carry out experiments; thus they perceive sciences as dull, theoretical and abstract. They fail to relate what they are taught with its application in the real world. Science will remain an abstract pursuit to learners so long as they are not exposed to its real application in their daily lives. Technology will never be appropriate if students are not afforded means of contextualizing it – this should earnestly begin in our laboratories ... (Prof. Shem Wandiga, Professor of Chemistry at the University of Nairobi, quoted in UNESCO 2014)

The technologies of engineering design will help address this challenge as students tinker with designs from their imaginations and see their designs come into life in the real world and not stuck in the abstract. The technologies will provide a great boost in stimulating interest in engineering among the school children and the youth.

With the current state of engineering education, Kenya is unlikely to meet its needs for 30,000 more engineers by 2030. Kenya should however enhance the capabilities of its engineers by using available engineering design technologies. Not only are these technologies labour saving; they are also skills saving (Edquist and Jacobson 1988) The skills deficit can be mitigated with the extensive deployment of these technologies. Engineering technologists and technicians could do the job of highly qualified senior engineers if they are equipped with CAD and other software skills. One engineer, with the help of various engineering applications, can do a job that until now was done by three or more engineers. It is, therefore, with the dual purpose of stimulating engineering education and meeting the skills deficit in multiple ways that Kenya should accelerate its uptake of engineering design technologies and overhaul its engineering design education.

Conclusion

The opportunities afforded by engineering design technologies need to be seized effectively to change the existing state of engineering design and to chart its future trajectory. Kenya and the wider Africa need to make the most of their latecomer's advantage by tapping into the latest technologies that are not only labour saving but also skills saving. These technologies have the potential to compensate for the shortage of a skilled workforce in a country. The lowering cost of software and hardware, and the wide availability of learning resources, are added advantages for Kenya where entrepreneurial talent is evident as exemplified by such innovative products as M-PESA and Ushahidi. There is reason for optimism that engineering design in Kenya will become a globally competitive activity. What is helpful in Kenya is the fact that Kenya is 'one of the most liberal and investor-friendly business environments on the continent', according to Adan Mohamed, Kenya's former Minister of Industrialisation and Enterprise Development.

Acknowledgement

The author acknowledges the research assistance provided by Zainab Meija Mohamed and Derick Muchele.

Disclosure statement

No potential conflict of interest was reported by the author.

Notes

1. The shortfall of 30,000 engineers is calculated on the basis of the ratio of at least 500 engineers and engineering technologists to 1 million people of the population needed for industrial take-off. By 2030, Kenya's population is projected to hit the 60 million mark. Estimated using this ratio, Kenya needs to have trained and qualified some 30,000 engineers and engineering technologists with a breakdown of at least 7500 engineers and 22,500 engineering technologists, in addition to 90,000 engineering technicians, and 450,000 craftspersons or artisans who will be required (Some 2013).
2. The basic skills in sketching, drawing and rendering for styling and form must be taught using pencil and paper, and that is a fundamental reason to continue teaching engineering design manually. This, however, should be limited to introductory and preparatory courses.

References

Antonelli, Paola. 2008. *Design and the Elastic Mind*. New York: Museum of Modern Art (MOMA). https://www.moma.org/interactives/exhibitions/2008/elasticmind/

Appiagyei Kwadwo Amoako, and Jonas Addai-Mensah. 2006. "Engineering Education and Graduate Attributes in Post-Modern West African Nation – Ghana: Technical & Social Drivers and Challenges." Paper presented to the 3rd African regional conference on engineering education, 'Engineering Education for Sustainable Development', University of Pretoria, Pretoria, South Africa, September 26–27, 2006. Accessed April 18, 2014. http://www.aeea.co.za/i/i//ARCEE3.pdf.

Asby, M. F., and D. R. Jones. 1994. *Engineering Materials 2 – An Introduction to Microstructures, Processing & Design*. Oxford: Elsevier Science.

Bell, Martin. 2007. *Technological Learning and the Development of Production and Innovative Capacities in the Industry and Infrastructure Sectors of the Least Developed Countries: What Roles for ODA?* UNCTAD The Least Developed Countries Report 2007 Background Paper. http://unctad.org/sections/ldc_dir/docs/ldcr2007_Bell_en.pdf.

Derbyshire, A. 2010. *Mechanical Engineering*. 3rd ed. Amsterdam: Elsevier Linacre.

Edquist, Charles, and Staffan Jacobson. 1988. *Flexible Automation: The Global Diffusion of New Technology in the Engineering Industry*. Oxford: Basil Blackwell.

Eide, Arvid, Roland Jenison, Larry Northup, and Lane Mashaw. 2002. *Introduction to Engineering Design and Problem Solving*. 2nd ed. New York: McGraw-Hill.

Elliott, Chris. 2010. "Engineering as Synthesis – Doing right Things and Doing Things Right. In *Philosophy of Engineering*, Volume 1 of the proceedings of a series of seminars held at The Royal Academy of Engineering, 54–57. London: Royal Academy of Engineering. www.raeng.org.uk/philosophyofengineering.

Erastus, Kabando K., and Pu Wuchuan. 2014. "Flaws in the Current Building Code and Code Making Process in Kenya." *Civil and Environmental Research* 6 (5): 24–30.

Falade, Funso. 2006. "Engineering and Globalisation in Developing Countries: Nigeria a Case Study." Paper presented to the 3rd African regional conference on engineering education.

Jordaan, Pietman W. 2006. "Is the Current Engineering Drawing Curriculum at Tertiary Institutions Still Applicable?" Paper presented to the 3rd African regional conference on engineering education, 'engineering education for sustainable development', University of Pretoria, Pretoria, South Africa, September 26–27, 2006. Accessed April 18, 2014. http://www.aeea.co.za/i/i//ARCEE3.pdf.

Muindi, B. 2014. Daily Nation: *Engineering courses in Kenya public universities suspended over quality*. Accessed April 18, 2014. http://mobile.nation.co.ke/news/Engineering-courses-in-Kenya-public-universities-suspended-/-/1950946/2517024/-/format/xhtml/-/4hxhqs/-/index.html.

Omura, G., and B. C. Benton. 2012. *Mastering AutoCAD 2013 and AutoCAD LT 2013*. Indianapolis, IN: Wiley.

Onwuka, E. N. 2009. "Reshaping Engineering Education Curriculum to Accommodate the Current Needs of Nigeria." *Educational Research and Review* 4 (7): 334–339.

Oxford Business Group. 2014. *The Report Kenya 2014*. Nairobi: Oxford Business Group.

Pohit, G., and G. Ghosh. 2007. *Machine Drawing with AutoCAD*. New Delhi: Dorling Kindersley

Roessner, David. 2008. *National and Regional Economic Impacts of Engineering Research Centres: A Pilot Study*. Menlo Park: SRI International.

Royal Academy of Engineers. 2012. *Engineers for Africa: Identifying Engineering Capacity Needs in Sub-Saharan Africa*. London: Royal Academy of Engineers. Accessed April 18, 2014. http://www.raeng.org.uk/international/activities/pdf/RAEng_Africa_Summary_Report.pdf.

Some, D. Kimutai. 2013. "Recent Developments in Higher Education Sector." Paper presented at the 17th annual scientific conference, Safari Park Hotel, May 8, 2013. Accessed November 12, 2015. http://www.kapkenya.org/pdfs/KAP.pdf.

The Engineering Act No. 43 of 2011. 2012. Nairobi: Kenya Laws Report. National Council.

The Engineering Technologists and Technicians Bill. 2015. Nairobi: Kenya Laws Report. National Council.

Turnbull, John. 2010. "The Context and Nature of Engineering Design." In *Philosophy of Engineering*, Volume 1 of the proceedings of a series of seminars held at The Royal Academy of Engineering, 30–34. London: Royal Academy of Engineering. www.raeng.org.uk/philosophyofengineering.

UNESCO. 2010. *Engineering: Issues, Challenges and Opportunities for Development*. Paris: UNESCO Publishing.

UNESCO. 2014. "UNESCO Mentors Girls in STEM to Mark the Africa Engineering Week." Accessed November 12, 2015. http://www.unesco.org/new/fileadmin/MULTIMEDIA/FIELD/Nairobi/images/KenyaMentoringSTEMAfricaEngWeek.pdf.

Waruru, Maina. 2015. "Universities versus Engineering Board Row Ecalates." *University World News*, Issue No 380, 4 September 2015.

Student's perception of engineering design for competitiveness in Africa: The case of Tanzania, East Africa

Lawrence Joseph Kerefu and Juliana Zawadi Machuve

Engineering is vital for addressing basic human needs, improving the quality of life and creating opportunities for sustainable prosperity at local, regional, national and global levels. However, Africa faces a shortage of engineers arising from the declining interest and enrolment of young people in the relevant disciplines at higher education institutions. Africa is also flooded with numerous products from the five major emerging economies, Brazil, Russia, India, China and South Africa, creating challenges for local engineering solutions. In this sense, engineering design capacity practice needs to be reviewed for the purpose of creating the potential for the application of local competitive engineering capacity for growth. This suggests a need for the upgrading and integration of local engineering design capacity into the mainstream, beginning with the education system. Empirical evidence from Tanzania lends support to this proposition. Primary data from undergraduate students suggest a strong focus on curricular review of engineering design capacity. Secondary data from the engineering design functions practised by industries and engineering-based R&D institutions indicates the existence of a skills mismatch which needs to be addressed. This holds important policy implications; in particular, the tertiary education curriculum ought to be sensitive to local needs and knowledge.

Introduction

Developments in engineering design have been central to human advancement in the past few decades. Engineering and technology have changed the world we live in, contributing to appreciably longer life expectancy and improved quality of life. Engineering brings many benefits including housing, transport, communications and healthcare. However, these benefits are distributed unevenly throughout the world particularly, in Africa. Most of the people in Africa do not have easy access to clean drinking water, suitable sanitation, medical centres and education services. Tanzania as one of the Eastern Africa countries is also facing these socioeconomic challenges.

In addressing these challenges, several development goals have been set such as the Millennium Development Goals (MDGs), and the Sustainable Development Goals (SDGs). The target to provide primary education for all, calls for new schools and roads to be built; just as improving maternal healthcare requires better and more accessible facilities; and better pollution control requires friendly environmental sustainability, clean technology, and improvements in farming practices (ElMaraghy 2010). In order to achieve these targets, African countries must implement policies which stimulate technological development and achieve economic diversification through innovation (UNCTAD 2007). One of the core components of innovation is engineering design, which is also a crucial driver of economic development in any nation. It is for this reason that engineering design capacity building deserves our attention, and its contribution to economic development in Africa must be acknowledged fully.

The East Africa region is facing a shortage of engineering design capacity, and thus highly depends on imported products from emerging national economies, namely, Brazil, Russia, India, China and South Africa (BRICS). Engineering design is pre-requisite to manufacturing and, therefore, it is important for the region to build and nurture engineering design capacity so that local solutions can be developed within the region. Tanzania is one of the Eastern Africa countries facing a shortage of engineering design capacity and this paper uses Tanzania for illustration.

According to the Tanzania Commission for Universities (TCU), fewer students are enrolling in the universities to pursue disciplines in engineering than disciplines in social science. Furthermore, students enrolling in engineering do not favour the engineering design course. Previous research in Tanzania has not focused on students' experiences in engineering design education. It should be taken into account that students are the main stakeholders and the future implementers of engineering design; therefore, there is a need for a study such this one to obtain the views of the students themselves with regards to engineering design education.

The main objective of this study is to assess the students' perception of engineering design in Tanzania, in order to improve the production of qualified and competent engineering design graduates. The overarching question asked this paper is: '*What is the status of engineering design education in Tanzania as perceived by the students?*'

Literature review

Tanzania – country profile

The United Republic of Tanzania, where the study was conducted, is located in Eastern Africa. It is bordered by eight countries namely, Kenya and Uganda to the North; Rwanda, Burundi and the Democratic Republic of Congo to the West; and Zambia, Malawi and Mozambique to the South. The country's eastern border, which has a coastline of 1,424 km (United Republic of Tanzania 2015), lies on the Indian Ocean. The official capital city of Tanzania is Dodoma, which is located in the central part of the country. Dar es Salaam is the country's largest commercial capital and the major seaport for serving the country's landlocked neighbours.

According to the 2012 Population and Housing Census (PHC), the population of Tanzania is 44,928,923 and the projection by 2018 was 54,199,163 (United Republic of Tanzania 2012). Tanzania is one of the most diverse countries in Africa, consisting of about 125 ethnic groups. Swahili is the national language and is widely spoken while English is the official language for higher education, administration and business. Tanzania's main economic sectors include agriculture, industry (mining and quarrying, electricity and natural gas, manufacturing, water supply and construction) and services (telecommunications, financial and business, trade and tourism). Tanzania's economy has been heavily dependent on the agriculture sector, which accounts for 24.7% of its Gross Domestic Product (GDP) and employs 75–80% of the total workforce (ESRF 2013). However, over recent years, the contribution of the agriculture sector to the GDP has been declining, while the contribution of other sectors, namely the services and industry sectors, has been increasing (ESRF 2015).

Tanzania's education system consists of pre-primary and primary education (7 years); ordinary and advanced level secondary education (6 years); and tertiary education (3 or more years). According to 2012 PHC, the literacy rate in Tanzania is 78.1% for persons aged 15 and above (United Republic of Tanzania 2012).

Overview of engineering design education
The first studies on engineering design in the engineering curricula date back to the 1950s where, particularly in Germany, it was developed as a sub-discipline within the topic of machine elements. Since then, the main trend in engineering education has been to develop engineering

methodologies and practices building on science-based theories and methods. However, the demand for engineering skills has changed resulting in a huge number of specializations and intricate responsibilities that engineers have to handle (Jørgensen 2008).

According to Dym et al. (2005), 'design', from an engineering point of view, is a systematic, intelligent process in which designers generate, evaluate, and specify concepts for devices, systems, or processes whose form and function achieve clients' objectives or users' needs while satisfying a specified set of constraints. Wood et al. (2001) emphasize the unique nature of 'designing' and that 'design' distinguishes between an engineering education and a science education. Various studies show that design is a central element in the engineering curriculum; however, it has been a challenge for the students to learn, and harder still to teach, and the retention rate of students in engineering design has been poor too (Adams, Turns, and Atman 2003; Dym et al. 2005; Bernold, Spurlin, and Anson 2007). Furthermore, previous studies have determined a range of factors which affect engineering design education including curricula and pedagogical experiences, policies and institutions, and students' readiness to study engineering (Bernold, Spurlin, and Anson 2007; Mhilu, Ilemobade, and Olubambi 2008).

Engineering design education in Tanzania
The engineering education system in Tanzania requires Advanced Secondary School students with passes/credits in any three of the following subjects: Physics, Mathematics, Chemistry, Biology or Geography to qualify for admission to the universities. Admission to the universities is carried out by the Tanzania Commission for Universities (TCU). TCU was established on 1 July 2005, under the Universities Act (Chapter 346 of the Laws of Tanzania). It is a body corporate mandated to recognize, approve, register and accredit the universities which are operating in Tanzania and local or foreign university level programmes being offered by registered higher education institutions (United Republic of Tanzania 2011).

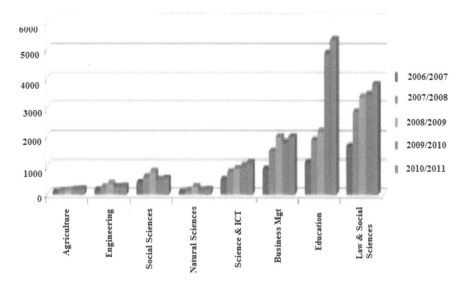

Figure 1: Student enrolment in university and university colleges by programme categories: 2006/2007–2010/2011.
Source: URT, TCU 2011

Figure 1 shows the enrolment trend in universities and universiy colleges by programme categories, from which it can be observed that the enrolment of engineering students is relatively low compared to social science programmes.

Since university education is more expensive than primary and secondary education, the government established a Higher Education Students' Loans Board (HESLB) by Act No. 9 of 2004 which became operational in July 2005. HESLB's objective is to provide loans for needy students who secure admission to accredited higher learning institutions, to pay for their education costs (United Republic of Tanzania 2010). The loan award criteria and conditions employed by HESLB are educational background, social economic status, parents' educational level, parents' occupation, parents' assets, and parents' lifestyle. However, in an attempt to encourage more students to pursue engineering courses, loan applicants for admission to higher education institutions to pursue such courses are given priority over non-engineering applicants.

Engineering training at university level in Tanzania takes four years before students can graduate. During the four years, students are required to attend industrial/practical training in each of the first three years for about eight (8) weeks and to conduct a final year project. Currently there are nine (9) universities and colleges in Tanzania which offer engineering courses graduating more than 1000 engineering students per year. The element of engineering design, which is the core subject of engineering training, is part of the curriculum in all disciplines of engineering. After engineering students graduate from universities they need to gain hands-on professional experience under close supervision by senior registered professional engineers. This professional experience for

graduate engineers is facilitated by the Tanzanian Engineers Registration Board (ERB), the government organ under the Ministry of Infrastructure Development.

ERB regulates the conduct of engineers and provides for their registration and related matters. ERB registered engineers at the level of professional engineers from 1968 to 1997 when it took on board young engineers as graduate engineers; in 2002 ERB introduced an internship programme for graduate engineers. ERB recognizes sixteen (16) disciplines of engineering specialization which have emerged from the three core engineering disciplines namely, mechanical, electrical and civil. the dominating discipline is civil engineering (46% of total registered engineers) as at June 2014, whereas the discipline with the smallest number of registered engineers is textile engineering (0.04%). Engineering design is featuring in all the disciplines of engineering specialization; however, the specializations do not develop the same type of approach. For example, in mechanical engineering, the basis is on the design of machines, in civil engineering, the focus is particularly on the design of roads and house construction, while in electronics the target is the design of functional artefacts of circuits (Jørgensen 2008). Table 1 shows the ERB's registration statistics of registered engineers by discipline and registration category.

Graduate engineering students are required by law/Act to register at ERB before practising as a graduate engineer. Afterwards, the graduate engineer is required to practise in the field/industry for three (3) years through the internship programme, known as the Structured Engineers Apprenticeship Programme (SEAP). SEAP intends to provide a smooth transition between academic training and professional practice. During this period, the ERB monitors and assesses how the graduate engineers practise engineering works/services

Table 1: Registration statics of registered engineers by discipline as at 30 June 2014.

Discipline	Category							Total	%
	GIE	GE	IE	PE	TPE	CE	TCE		
Civil	221	2695	145	2090	890	240	78	6230	46%
Mechanical/Industrial/Automobile	133	980	73	655	107	33	5	1975	14%
Electrical	145	1241	137	578	100	32	3	2236	16%
Electronics & Telecommunication	166	491	36	130	33	8	–	914	6%
Environmental	27	385	22	123	5	6	2	652	4.3%
Mining/Mineral Processing/Petroleum/Gas	–	311	2	99	57	3	1	483	3.5%
Aeronautical	–	2	–	6	–	–	–	8	0.1%
Agricultural/Irrigation	–	207	–	57	5	3	1	275	2.0%
Chemical & Process	2	388	2	97	13	3	–	504	3.8%
Marine	1	5	4	23	2	–	–	46	0.4%
Computer & IT	12	315	–	21	–	–	–	336	2.5%
Textile	–	2	–	2	1	–	–	5	0.04%
Geotechnical	–	20	–	7	8	3	1	39	0.3%
Food and Biochemical	–	27	–	–	–	–	–	27	0.2%
Electromechanical	–	82	–	10	2	2	–	96	0.7%
TOTAL	**707**	**7151**	**421**	**3898**	**1223**	**333**	**93**	**13826**	**100%**

Legend
GIE: Graduate Incorporated Engineers
GE: Graduate Engineers
IE: Incorporated Engineers
PE: Professional Engineers
TPE: Temporary Professional Engineers
CE: Consulting Engineers
TCE: Temporary Consulting Engineers
Source: Engineers Registration Board (ERB) 2015

Table 2: SEAP assessment criteria and rating in areas of professional practice.

Area of practice	Assessment criteria	Rating	Score
1. Field/Site/Workshop Competence	1.1 Maintenance practice	0–10	
	1.2 Ability to use theoretical and applied knowledge in independent practice	0–10	
	1.3 Problem diagnosis	0–5	
	1.4 Problem investigation and solving	0–5	
	1.5 Laboratory work or machine and workshop practice	0–5	
	1.6 Trouble shooting	0–5	
	Total	40	
2. Design Competence	2.1 Application of engineering standards	0–4	
	2.2 Innovativeness depicted	0–4	
	2.3 Balance between technical effectiveness	0–4	
	2.4 Design calculations and drawings	0–10	
	2.5 Specifications	0–4	
	2.6 Quantities and estimating	0–4	
	Total	30	
3. Management	3.1 Ability to make effective engineering decisions	0–5	
	3.2 Ability of innovative planning, design and management	0–4	
	3.3 Staff and Labour Management (Material, labour etc)	0–3	
	3.4 Programming and Estimating	0–3	
	3.5 Maintenance and Management	0–3	
	3.6 Costing and accounts	0–2	
	3.7 Quality assurance	0–2	
	3.8 Safety	0–3	
	3.9 Environmental issue	0–3	
	Total	30	
	GRAND TOTAL	**100**	

Source: ERB, 2015

by focusing on three main activities, namely: planning and design, field/site/workshop practice (e.g. installation, maintenance, production, etc.) and engineering management (e.g. supervision of people, projects, quality, finances, health and safety etc.). The assessment criteria and rating are shown in Table 2. SEAP provides detailed guidelines in which the element of engineering design is mandatory for all 16 disciplines. Upon successful completion of SEAP, the graduate engineer is qualified for registration as a professional engineer and is awarded a registration certificate. In 2011, ERB launched the Practicing Licence which every professional and consulting engineer requires, in addition to the registration certificate.

Factors and policies affecting engineering design education in Tanzania

Factors

It is envisaged that although the engineering training system in Tanzania is well structured, several factors have the potential to influence engineering design capacity. These factors include:

(i) The educational pipeline from university to field/work practice (enrolment, teaching methods, student's attitude, labour market, few industries to practice): Fewer students enrol to pursue engineering disciplines than other disciplines such as business, law, sociology, etc. It is perceived that labour market demand and students' attitude are the reasons for this imbalance. Nowadays, students are much more attuned to entertainment issues than educational ones, and therefore there is a need for new motivating teaching and learning methods (e.g. problem-based learning approach). There are only a few manufacturing industries in the country and therefore practical training opportunities for students to practise engineering design is

limited. Furthermore, these few manufacturing industries do not have proper design offices/design departments. Despite these limitations, there is a need to conduct research on the engineering design educational pipeline to identify strategies for improving the transition from university to field/work practice.

(ii) Weak link between universities, industries and R&D institutions: Universities cooperate with industries and R&D institutions for practical training of students. There is linkage with industry through consultancy projects involving staff members whose experience in these projects enhances their teaching capability. The linkages with R&D institutions are normally engineering design oriented. However, these linkages are weak due to poor coordination and limited resources, but there is a need to further investigate and identify the weaknesses in the linkages.

(iii) Imbalance in the engineering cadre (engineers, technicians, artisans): Over the past few years in Tanzania, there has been a growing tendency of upgrading existing technical colleges into universities. This had led to a reduced number of technicians and artisans that has caused an imbalance of in the availability of engineers, technicians and artisans. According to the International Labour Organization (ILO), the standard skills-mix ratio among engineers, technicians and skilled artisans is 1:5:25. However, in Tanzania the current ratio is 1:2:14 (DIT 2014). Therefore, there is a need to develop the technician and artisan cadres who assist engineers in conducting engineering design activities.

(iv) Field/work practice (poor motivation, few proper design office/design departments in industry, limitations of technologies, ICT, evolving/new technologies): Effective engineering education needs to integrate general

education, formal practical training and professional experience in the field. However, in Tanzania there is poor motivation to practise engineering design in the field; there are few proper design offices/design departments in the field/industries; and the existing technologies in industries and R&D institutions need to be upgraded to new technologies, including the use of ICT.

(v) Shortage of funding: Engineering studies require more funds than non-engineering studies because they need to cater for equipment, laboratories, machines, etc. In Tanzania, the engineering universities have insufficient funds for carrying out operations, thus affecting the quality of engineering design education offered.

(vi) Industries and engineering-based R&D institutions in Tanzania: In Tanzania, industries and engineering-based R&D institutions are the main employers of engineering graduates from local higher learning institutions. They provide practical training for engineering students, whereby the students get the opportunity to improve their engineering design competencies. On the other hand, the government of Tanzania recognizes the leading role of the industrial sector in the transformation of Tanzania's economy. Despite the industries and R&D being linked to economic growth, manufacturing has a key role in transforming the economic structure. In Tanzania, manufacturing cuts across various sectors including mineral processing, metal processing, food processing, plastics, petroleum/petrochemical, automobile, energy, construction, chemicals, wood, paper, drinks, building materials, textiles, etc.

Most of the R&D institutions in Tanzania are non-corporate, created and supported by the government to provide domestic industries a technological edge . Engineering-based R&D institutions introduced the applications of Computer-Aided-Design (CAD) way back in the 1990s, providing users the advantages inherent in managing general data for engineering design. Despite an increase in the use of ICT tools in design processes, designers still rely on personal exchanges and visual communication in solving problems and carrying out innovations. In order to attain the desired design and engineering capability, training for advanced Computer-Aided Design and Manufacturing (CAD/CAM) and high precision production have to be conducted.

Policies /regulatory framework
In Tanzania there are several policies which influence engineering design practice and capacity. They include: Education and Training Policy, National Research and Development (R&D) policy, Sustainable Industries Development Policy and National Investment Promotion Policy. In addition, there are regulatory frameworks that have the potential to influence engineering design capacity, such as Engineers Registration Board (ERB), Tanzania Commission for Science and Technology (COSTECH), and Tanzania Commission for Universities (TCU). There is a need for further study of these policies and regulatory frameworks to recommend how they can be used effectively to promote engineering design capacity and practice.

Methodology
The research methodology of this study involved positivism research philosophy, which depends on quantifiable observations, focusing on facts and viewing the world as external. A deductive approach was used, and the study adopted a descriptive research design, which aims at describing characteristics of individuals, groups or situations. The time horizon for the study was cross-sectional. The data collection was conducted using semi-structured questionnaires and interview guides. Thereafter the data were analyzed by descriptive statistics using the SPSS software package. The study's methodology is summarized in Figure 2.

Data collection and analysis
Data collection
Data collection was carried out using semi-structured questionnaires, which were physically sent to four higher

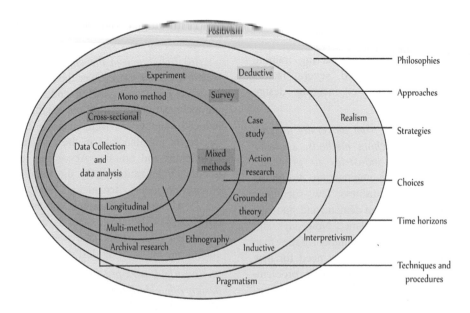

Figure 2: The research onion (Saunders 2009).

learning institutions in Tanzania that offer engineering degree programmes. Before using the questionnaire in the full-scale study, a pilot survey was carried out. The aim of the pilot was to pre-test the questionnaire in order to improve its content and increase the validity of the study's concepts as well as to detect any problems arising from administering the questionnaire. The questionnaires were distributed to engineering students studying engineering design for the purpose of obtaining information about their perceptions of the status of engineering design education. The areas which were considered included the students' personal background, their education background, their perceptions, learning and teaching of engineering design.

Data analysis and interpretation of results

Personal background

(i) Gender distribution: The gender distribution for students pursuing engineering design shows that female students make up 30% (Figure 3) which is below the government target of 50%. However, the trend towards balance is slowly increasing compared with previous years.

(ii) Age distribution: The age distribution of engineering design students shows that most of them (75.6%) are aged between 20–25 which is a suitable age for engineering since it requires creativity and flexibility for site work (Figure 4).

(iii) Surveyed higher learning institutions: Four (4) higher learning institutions which offer engineering design subjects were surveyed. This is only 10.8% of the 37 existing fully fledged higher learning institutions in Tanzania.

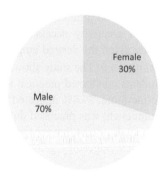

Figure 3: Students distribution by sex.

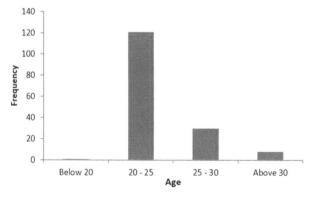

Figure 4: Students distribution by age.

(iv) Engineering degree programme: Engineering design students study under various degree programmes (shown in Figure 5) all of which have components of engineering design. Civil engineering has the largest number of students (36.9%), followed by mechanical engineering (13.8%). In this survey, engineering design featured primarily in civil engineering, which can be attributed to the availability of employment opportunities in the construction industry in Tanzania.

(v) Year of study: Data was collected from second to fourth year undergraduate engineering students (Figure 6) because the engineering design component is mainly taught and practised from the second year onwards.

Education background

(i) Secondary school subjects and facilities: In Tanzania most private schools are better equipped with teaching facilities such as laboratories, libraries, and competent teachers than government schools. This study shows that 28.1% of the engineering students were from private schools, indicating that about 28.1% of these students have a good education background. The majority of the students (73.1%) indicated that they had good quality teachers in Mathematics and Science and 60.6% of the students indicated that they had well-equipped laboratories which are essential for conducting practical work in science (Figure 7). The government is taking measures to address the shortage of teaching facilities, particularly through the construction of laboratories for science subjects in public secondary schools.

(ii) Entry grade into university: The basic foundation subjects for pursuing engineering design courses are science subjects including: Physics, Mathematics, Chemistry, Biology and Geography. For the surveyed students, it was observed that their entry grades into university were above average: Mathematics (67.5%), Chemistry (59.4%), Physics (56.2%), Biology (20.6%) and Geography (23.1%). It was also noted that most students admitted to engineering studies did not take Biology or Geography in their Advanced Level secondary studies.

(iii) Motivation to choose an engineering career: During the year 2010/2011 the TCU report shows that the total number of students admitted to universities was 135,367 of which 3001 were enrolled for engineering courses, which is only 2.22% of the entire national cohort. This calls for strategies to motivate students to study science subjects, in particular subjects related to engineering.

This study shows that 75% of the surveyed engineering students were inspired to choose an engineering career due to their love for and interest in engineering. Other reasons such as to obtain a good job, prestige, good salary and ease of obtaining a study loan when pursuing an engineering degree did not seem of much importance to the respondents. However, additional motivation reasons mentioned by students included: life goals, to complement their career choice, to get knowledge, to use engineering skills to develop a business that would provide employment in villages, to develop a career, and self-employment.

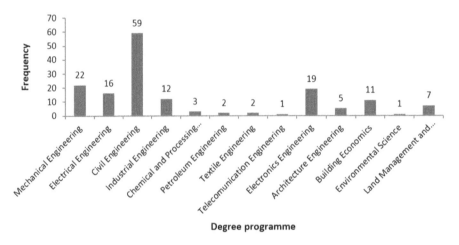

Figure 5: Students distribution by degree programme.

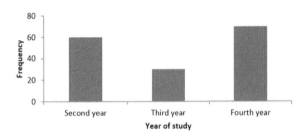

Figure 6: Students distribution by year of study.

Engineering design subject

(i) Engineering design content in curriculum and its importance: The study shows that engineering design content is part of the curriculum as one of the core subjects. The engineering universities in Tanzania have taken note of the importance of engineering design and thus consider it a mandatory and not an optional course.

(ii) Rating of the importance of engineering design: The study observed that 50% of the surveyed students indicated having a very high interest in studying engineering design. The adequacy of the learning materials was rated high by 36.9% of the students. The appropriateness of teaching methods was rated high by 30.6% of the students, while 33.0% rated the quality of the teaching staff high, and 30.6% rated the availability of equipped

laboratories high. Similarly, 44.4% rated the relevance of the subject to jobs and future employment interest high, and 51.2% rated their overall performance in engineering design subject high.

Although the students rated these factors as high, the existing situation shows that the overall curriculum has not included new technologies such as 3D printing and nanotechnologies which are design vehicles in engineering design subjects.

(iii) Assessment: The study found that 51.9% of the surveyed students indicated that the methods of assessing engineering is by creating real designs, while 45% indicated that the assessment is by re-working existing designs. Other methods of assessment which were mentioned by the students include: doing assignments without real design, doing written tests and examinations. These additional assessment methods mentioned show that they are not based on problem-based learning (PBL) and the element of creativity necessary in engineering design is not being given the desired emphasis.

(iv) Practical training: The study shows that 98.1% of the surveyed students conducted practical training during their studies. Furthermore, 79.4% indicated that the engineering design component was practised during their practical training. The study shows that the students, to a large extent, used the knowledge they gained from university

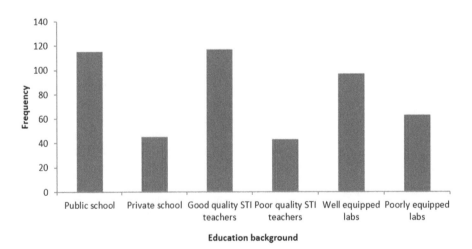

Figure 7: Students' education background.

during their practical training, indicating that the engineering design training at the university is relevant.

(v) Encouraging other students to study engineering design: The study shows that majority (97.5%) of the students will encourage other students to study engineering design. The reasons they mentioned include: increasing design skills; can solve the problem of unemployment; design is a major part of engineering as it enables a person to solve social and practical problems in surrounding societies and thus it is important for the wellbeing of society at large; engineering design helps to broaden students' knowledge about how to discover different products; through studying engineering design students can improve local technologies and decrease external dependence; the subject is very important in all sectors – e.g. in building construction where they can design and plan a variety of electrical systems and in agriculture where they can design irrigation systems; it is an interesting subject with a lot visuals; it involves practical training and hence helps transform knowledge to hands-on experience; it touches important aspects of daily life; it has real-life application because through design one can innovate components to complete various tasks; it encourages students to think creatively about new ideas to integrate with techniques for solving different problems in our societies; it is an interesting course since it is a way to make the youth plan for self-employment; there is a need to increase the number of engineers who can carry out design projects; there is demand since most companies need people with design skills and the market needs new products; it makes it easier to access a wide range of high-paying job opportunities; the subject makes students critical thinkers and to practise their knowledge well; it trains them to be competent and confident and to become job creators rather than job seekers by using the skills obtained through engineering design; because there is a shortage of professionals in the engineering field and in order to get more indigenous design in our country; it can aid students to be innovative, to create new designs and carry out modifications on existing design failures; it can make them aware of the technology revolution which is always changing globally

(vi) Intention after graduation: The study shows that 73.8% of the surveyed students do not intend to pursue higher degrees in engineering design after graduation. Of the surveyed students, 46.9% indicated that they would create their own jobs, 34.4% indicated that they would join existing companies while 76.95% indicated that they would not pursue becoming consultants.

Other things which the students mentioned that they would do after graduation were to engage in large-scale entrepreneurship; create a company; study other areas like architecture or building economy; and self-employment, to become an entrepreneur in engineering.

The comments by the students in sections (v) and (vi) above show that there is no promising future for engineering design training at postgraduate level. Furthermore, the comments show that the mindset of the students calls for entrepreneurship and innovation modules to be included in the engineering curriculum.

(vii) Engineering design practice in industries and R&D institutions: The existing industries in Tanzania have limited engineering design activities. Most of them do not even have engineering design offices/departments. Therefore, engineering graduates do not get good opportunities to practise engineering design in a work context. And the few industries that possess engineering design sections/offices are not flexible enough to accommodate rapid technology changes. Further, engineering-based R&D institutions which carry out engineering design functions do not have advanced engineering design tools such as 3D printers and engineering design software to enable them to fast track the engineering design process. Industries and R&D institutions argue that they do not receive engineering design graduates who possess the skills and expertise needed to function effectively in their workplace.

From this situation, and from the students rating of the importance of engineering design (section (ii) above), which shows a skills mismatch of 44.4%, both industries and R&D institutions are required to improve engineering design capabilities to enhance their competitiveness in the face of increasing market demand. Products that R&D institutions design should be market oriented, targeting real-life problems, and involving all stakeholders. There should be linkages between industries and R&D institutions for the purpose of sharing their requirements. Furthermore, they should create incentive mechanisms to motivate engineering design staff. The university curriculum should be reviewed from time to time, in collaboration with strategic stakeholders, to address the requirements of industries and R&D institutions.

(viii) Students' suggestions: The perceptions of the students on the issue of engineering design were further obtained from their own suggestions, such as: there should a change in the way engineering design is taught to accommodate the development of modern technologies; there is a need to have both theory and practical training, but there should be much more practical training so that they are sufficiently competent when they graduate; the government should support young engineers, especially undergraduates, and motivate engineers rather than focusing so much on politicians, provide experts who teach engineering design with good incentives; improve modes of teaching and tools, improve laboratories and equipment; students should conduct study visits to different industries; to learn best practices of engineering design from developed countries by engaging in modern technologies such as automation systems; the government should provide students who graduate in engineering design with loans so they can create their own jobs and innovations; encourage students to re-work existing designs and then in future they can create real designs; the engineering design syllabus should be reviewed so that it becomes easier to teach; advise students to do extra practise in engineering design to gain more experience and skills; advise institutions that engineering design should be taught from the first year so that students can have a good foundation; local consulting companies should be given higher priority in tenders rather than giving them only minor design works; to create a system which can enable all engineering universities to meet together and

share ideas about the engineering field and training; do research in villages to find opportunities to apply engineering design skills; support from the private sector and government budget allocations for engineering design; young students should be motivated and inspired from their basic educational levels to pursue engineering design.

Conclusion and recommendations
This study was conducted for the purpose of assessing students' perception of engineering design education. Primary data from students were collected to provide a general picture of the current situation of engineering design education. It was found that students are facing challenges in pursuing engineering design due to various problems embedded in their training. The major ones include teaching methodologies, teaching facilities, and a skills mismatch resulting from the demand for engineering design capabilities by industries and R&D institutions. In order for students to attain the desired design and engineering capability, universities, government, industries and R&D institutions should jointly review the engineering design education curriculum. Similarly, R&D institutions and industries need to update their facilities with advanced technologies, including design and manufacturing software packages for sustainable competitiveness.

This research is limited to Tanzania and thus cannot be used exhaustively for the generalization of the status of engineering capacity in the region. Due to the fact that engineering design was found not to be broadly practised in the field, the study faced the limitations of a small sample size and the use of cross-sectional data due to the absence of and/or limited longitudinal data. It is recommended that further research is conducted in other East Africa regions to appraise the status of students' perception of engineering design and related engineering firms, R&D institutions, and foreign firms in terms of the utilization of local engineering design capabilities and on how indigenous knowledge is used for promoting engineering design and technology transfer.

Disclosure statement
No potential conflict of interest was reported by the authors.

References
Adams, R. S., J. Turns, and C. J. Atman. 2003. "Educating Effective Engineering Designers: The Role of Reflective Practice." *Design Studies* 24 (3): 275–294.
Bernold, L. E., J. E. Spurlin, and C. M. Anson. 2007. "Understanding our Students: A Longitudinal-Study of Success and Failure in Engineering with Implications for Increased Retention." *Journal of Engineering Education* 96 (3): 263–274.
Dar es Salaam Institute of Technology (DIT). 2014. *Improving Skills Training for Employment Program (ISTEP)*.
Dym, C. L., A. M. Agogino, O. Eris, D. D. Frey, and L. J. Leifer. 2005. "Engineering Design Thinking, Teaching, and Learning." *Journal of Engineering Education* 94 (1): 103–120.
Economic and Social Research Foundation (ESRF). 2013. Tanzania Annual Economic Review For 2013.
Economic and Social Research Foundation (ESRF). 2015. Quarterly Economic Review, 15(2), April-June, 2015.
ElMaraghy, H. 2010. "*Future Trends in Engineering Education and Research*." The 8th global Conference on sustainable manufacturing, Abu Dhabi University, Khalifa City.
Engineers Registration Board (ERB). 2015. Registration Information. Available at https://www.erb.go.tz/.
Jørgensen, U. 2008. "Engineering Design in Theory, Teaching, and Practice: Contemporary Challenges to the Construction of Design Synthesis and Methods." In *INES Workshop*. INES website.
Mhilu, C. F., A. A. Ilemobade, and P. A. Olubambi. 2008. *Preliminary Study: Engineering Education Evaluation*.
Saunders, L. T. 2009. *Research Methods for Business Students*. 5th ed. Harlow: Pearson Education Limited.
United Nations Conference on Trade and Development (UNCTAD). 2007. *The Least Developed Countries Report 2007: National Policies to Promote Technological Learning and Innovation*. New York and Geneva: UNCTAD.
United Republic of Tanzania (URT). 2010. Higher Education Students' Loans Board (HESLB). Available at http://www.heslb.go.tz.
United Republic of Tanzania (URT). 2012. *Report on 2012 Population and Housing Census (PHC)*. Dar es Salaam: National Bureau of Statistics.
United Republic of Tanzania (URT), Tanzania Commission for Universities (TCU). 2011. *Enrolment 2005 - 2015*. Available at http://tcu.go.tz/images/documents/Enrolment_2005_2015.pdf.
United Republic of Tanzanian (URT), Engineers Registration Board (ERB). 2015. *Registration Information*. Available at https://www.erb.go.tz/.
Wood, K. L., D. Jensen, J. Bezdek, and K. N. Otto. 2001. "Reverse Engineering and Redesign: Courses to incrementally and Systematically Teach Design." *Journal of Engineering Education* 90 (3): 363–374.

Barriers hindering biomimicry adoption and application in the construction industry

Olusegun A. Oguntona and Clinton O. Aigbavboa

Attaining the goal of sustainability in the construction industry is a demanding task that comes with numerous challenges and complexities. To overcome these, biomimicry, as the study and emulation of nature's features, processes and systems for solving diverse human issues offers enormous potential. This research paper sets out to identify and present what constitutes the hindrances to biomimicry adoption and application in the construction industry (CI). The exploratory factor analysis (EFA) technique was employed to attain the aim of the research study. Out of the 120 questionnaires administered to construction professionals, 104 were completed and returned to establish their perception of the barriers to employing biomimicry in the CI. The result of the data analysis established four underlying barriers in order of their significance, namely *information and technology*, *risk and cost*, *knowledge*, and *regulations*. This methodical modus operandi towards comprehending the taxonomy of the barriers to adopting biomimicry in the CI is imperative for proffering an effective and efficient solution.

Introduction

The construction industry (CI) is closely linked to urbanization owing to the numerous developments associated with it (Shi et al. 2014). The purveying of essential infrastructure such as rail, roads, buildings (industrial, residential and commercial), bridges, plants for energy production, transmission and distribution, and wastewater and water treatment plants are all direct impacts of the industry (Pearce and Ahn 2012). These infrastructures have benefitted humanity and significantly contributed to the economic development of nations, especially the developing ones (Azis et al. 2012). With these feats amongst others, the industry is positioned and known to play an essential role in ameliorating the standard of human life and in satisfying the demands of the populace in question (Tam, Tam, and Tsui 2004). Governments' use of construction investments as a tool for stabilizing the economy is a validation that the industry performs a vital part in the national development strategy of many nations (Chang and Pheng 2011). The existence of a correlation between the economic upturn of developing nations and the CI has been identified, with the focus on infrastructural developments and industrialization. Thus, this relationship is seen to be a causal connection, which suggests that investments in infrastructure do lead economic growth by improving the efficiency and the capacity of the economy (Giang and Pheng 2011). The strong connection that exists between the CI and other sectors is a pointer to the industry's ability to stimulate the growth and development of any nation. The CI is not only an indispensable fraction of the modernization process, but its labour-intensive characteristics makes it also a mechanism for employment creation in developing nations (Ramsaran and Hosein 2006). This attribute of the industry is what has led developing nations to heavily invest in infrastructure with the sole focus on their economic growth and development while focusing less on the environmental impact.

Buildings and other infrastructures, which are upshots of the CI, have a protracted effect on the human and natural environment as they are incessantly responsible for the emission of a significant amount of pollution (Wang 2014). In addition to air, water and land pollution, the CI produces significant amounts of waste from the production and use of materials. Based on the appraisal of generated material waste on an identified Dutch construction project site, Azis et al. (2012) reported that materials wasted on the construction site is relatively excessive and amounts to 9% by weight of the total materials purchased. Also, waste is generated through construction and demolition, retrofitting and renovation works which buildings generally undergo in their lifecycle. Since most of the conventional materials utilized in the CI are not reusable or recyclable, they end up polluting the ecosystem and the human environment. In summary, the activities of the CI are only geared towards the economy, thus failing to holistically address and consider the environmental protection and social equity goals of sustainability.

Considering its social and environmental impact, Shi, Zuo, and Zillante (2012) identified the CI as an essential sector in achieving sustainability. With the continuously growing massive infrastructural developments around the globe, implementing sustainable interventions now rather than changing things afterwards is imperative. Three forces are identified by Kibert (2013) to be propelling the shift of the industry towards the sustainability path. Firstly, the pressures experienced by the continuous increase in demand for natural resources, with high prices and shortages of materials as the resultant effect. Secondly, the accelerated and massive destruction of planetary ecosystems and biodiversity, a significant increase in population and consumption, alteration of biogeochemical cycles, all resulting in the threat of global warming, depletion of marine life, deforestation, and desertification, amongst many others. Thirdly, the transformation movement coupled with the various approaches adopted in agriculture, tourism, manufacturing, medicine and the public sector, amongst others, towards greening their activities. These have led to the growing recognition and importance of adopting the principles of sustainability in the CI (Plank 2005). In seeking solutions to the issues of sustainability, the world has seen a multidisciplinary collaboration and cooperation resulting in different efficient and effective trends and approaches in the CI.

Sustainability trends in the construction industry

The efforts and drive of professionals and stakeholders in the CI have culminated in the birth of different paths and concepts to realizing the sustainability agenda of the CI. These are believed to possess considerable potential in shaping the future of the built environment. Some of these trends address the three pillars of sustainability in a holistic context while others concentrate on one or two parts. Examples of these novel concepts aimed at achieving sustainable objectives in the CI include Factor 4 and Factor 10; the Natural step; precautionary principle; biophilia; eco-efficiency; ecological economics; construction ecology; ecological footprint; industrial ecology; climate change and passive survivability; life-cycle assessment; life-cycle costing; lean construction; value engineering; industrialized building system; building information modelling; nanotechnology; design for the environment; cradle-to-cradle; and biomimicry (Pearce and Ahn 2012; Hussin, Rahman, and Memon 2013; Kibert 2013). Of the listed approaches, biomimicry is a relatively new field which encourages the teaming up of biologists with designers and professionals from various fields (industrial design, medical science, material science, architecture, and interior design) to study and emulate strategies and materials in nature, and thereafter ensure the knowledge transfer to the built environment. As affirmed by Benyus (2011), the idea behind biomimicry is prompted by the fact that the natural world has managed to exist for over 3.8 billion years with biota as archetypes that create opportunities rather than waste, manufacture without going through the process of heating, beating, and treating; and solar-powered ecosystems. Also, their designs have been found to be sustainable, effective, and attractive. Rather than engaging in nature exploitation, biomimicry considers and focuses on identifying and integrating propositions that are primarily sustainable and responsive to the earth's capacity (Goss 2009).

Historical antecedent of biomimicry

Practising and applying the knowledge of natural systems in human society is not eccentric. However, the term that describes it is relatively new. There is a growing inquisitiveness about the natural world around us due to the continuing quest for advancement and improvement in science and technology. Historically, early people relied solely on nature for existence and survival, coupled with records of native innovations as a result of human observation of nature and natural phenomenon. These include innovations in shelter architecture; medical sciences; weapons and defence (including alarm systems, drones, armour, and sensors); agriculture (animal husbandry and food production); and constituting processes of

manufacturing (Murr 2015). Early scientists and innovators have gathered invaluable information about the function, efficiency and sustainable exploitation of resources by mere examination as well as a comprehensive consideration of nature. Nature exploration has also revealed a robust system, the execution of which is as near perfection as it can be, whereby nature dispenses an inexhaustible fount of ingenuity for diverse fields of human endeavour. Every natural organism has been discovered to be idiosyncratic and wholly tailored to its environment (El-Zeiny 2012). Therefore, the natural world revamps and supports itself over the years by fulfilling its personal needs and proffering environmentally responsive solutions to every challenge it faces.

Jack Steele of the United States Air Force coined the word 'bionics' in the year 1960. He described the word as the science of systems which possesses some operations copied from nature, its characteristics or its operational strategies (Vincent et al. 2006). The word 'biomimetics' was conceived by Dr Otto H. Schmitt, a professor at the University of Minnesota from 1949 to 1983 (Minnesota Inventors Hall of Fame 2016). It was used for the first time in 1969 as the subject of a paper presented by Dr Schmitt at the 3rd International Biophysics Congress in Boston. In his doctoral research, he strived to create a device that in its entirety emulated the electrical action of a nerve. Biomimicry became known in the year 1997 with the publication of *Biomimicry: Innovation Inspired by Nature* authored by Janine M. Benyus. Benyus, globally acknowledged as the originator of this unconventional branch of study which her book helped to spread, is a biologist, and co-founded the Biomimicry Guild, (Goss 2009).

Definition of biomimicry

Throughout the literature, multiple terms are used to illustrate this convention of learning and drawing inspiration from nature. Aziz and El Sherif (2015) contend that no difference exists between 'biomimicry' and 'biomimetics'. Other terms like biognosis, bioinspiration, bio-inspired design, nature-inspired design, bioanalogous design, bionics, and biomimesis are often utilized interchangeably with biomimetics and biomimicry (Vincent et al. 2006; Shu et al. 2011; Gamage and Hyde 2012). Emerging from the merger of two Greek words, *bios* – life – and *mī'mēsis* – imitation, biomimicry can be inferred to mean 'life imitation' (Pronk, Blacha, and Bots 2008; De Pauw et al. 2010; Arnarson 2011; Gamage and Hyde 2012; Murr 2015). Benyus interprets biomimicry to be the pursuit of humankind in exploring nature's amazing attributes (self-sustaining ecosystems, natural selection, self-assembly, self-repair and photosynthesis among many others) and then emulating these systems and blueprints to deliver sustainable remedies to their own problems (Benyus 1997). El Din, Abdou, and El Gawad (2016) defined biomimicry as the exploration of shapes, systems, elements, processes, and strategies operational in the natural world with the prospects of solving human problems. According to Badarnah and Kadri (2015), it is described as proffering solutions to human issues by imitating the strategies, principles, and mechanisms inherent in nature. In summary, it is the process of creating sustainable designs through the studying and conscious imitation of extracted ideas from the attributes, operations, and ecosystems exhibited in the natural environment (Benyus 2011; Singh and Nayyar 2015).

The term biomimicry is judged by several researchers to be some aspect of biological science: the scientific knowledge within the field only suffice as a channel for studying and emulating nature (Marshall 2013). Biomimicry brings about a transition from the traditional norm of nature exploitation to learning from its structures, operations, and blueprints (Benyus 2011). Therefore, the potential to proffer sustainable resolutions to the various problems encountered by present-day humanity emanates from the proposition that the natural world has developed exceptionally coherent systems and strategies (Hargroves and Smith 2006). Biomimicry proponents are convinced that the CI should learn from the efficient research and development (R&D) laboratory in operation for the past 3.8 billion years of earth's existence. Here, 10 to 30 million floras and faunas have mastered all humans yearn to achieve, without disrupting the communal future of successive generations or engaging in environmental pollution (Strategic Direction 2008). According to Nychka and Chen (2012), nature has been unearthed as a robust fount of knowledge, engendering the progression and innovation of novel remedies to present-day human issues through the implementation of biomimicry. The majority of these issues (problems) are those initiated by the uncontrolled increase in exploitation of natural resources and industrialization globally (Rinaldi 2007). It is, therefore, believed that innovations, ideas, and solutions inspired by nature are now the panacea to the challenges facing humanity.

Biomimicry potentials and accomplishments

Biomimicry is achieving striking importance as a global phenomenon for eco-friendly and sustainable developments characterized by its potential to trigger ingenious and innovative solutions (Zari 2007; Gamage and Hyde 2012). Biomimicry can be deployed as a strategy to increase the sustainability of what already exists (Zari 2007), while new technological solutions and inventions can also be achieved through its application (Okuyucu 2015). Since it became popularized, biomimicry has provided several exceptional innovations in the domains of waste management and energy engineering which requires multiple-scale efficiency advancements. For instance, the Biomimicry Guild has assisted major multinational corporations and companies in applying nature's designs to create sustainable solutions and innovations (Benyus 2011). A few of these corporations include global architects HOK, Procter and Gamble, Nike and Interface, a carpet company, among many others. These examples have showcased the capability of biomimicry to develop efficient and high-performance products with sustainable characteristics (Oguntona and Aigbavboa 2016).

However, several factors are impeding the promotion, application, and performance of biomimicry in the CI. Gamage and Hyde (2012) identify five barriers to biomimicry adoption, namely: barriers of environmental principles and policies (non-implementation of biomimicry principles); language barrier (lack of understanding of approaches); barriers due to ecosystem complexities (inability to understand the strategies and processes exhibited in nature); integration barrier (inability to integrate the knowledge of biomimicry); and conceptualization barrier (inability to interpret biomimicry principles). As a global multidisciplinary phenomenon and an emerging sustainability concept in the CI, biomimicry is confronted with similar barriers facing sustainable construction. (Table 1).

Research methodology

Zari (2008) identified the interdependent approach of stakeholders and knowledge sharing, and different economic and legal frameworks as significant barriers to the acceptance and application of biomimicry in the built environment. The approach suggests that stakeholders and professionals in the built environment have essential responsibilities in ensuring biomimicry achieves its overarching goal of sustainability. To this end, practising biomimicry specialists and construction professionals in South Africa were sampled as they are predominantly responsible for the transition of the industry from the traditional ways towards the global sustainability paradigm. As significant players and green agents, their perceptions should help stakeholders and professionals in other countries to identify and understand the impediments to the adoption and application of biomimicry. The targeted respondents consist of practising biomimicry and construction professionals in the South African construction industry (SACI), all registered and active members of their various professional bodies. One-hundred-and-twenty (120) structured questionnaires were administered to extract their perceived extent to which the 19 variables listed hinder biomimicry adoption and implementation. One-hundred-and-four (104) duly completed questionnaires were retrieved which represents an 87% response rate. The sampled respondents were requested to specify their response on a 5-point Likert scale (not a barrier – 1, slight barrier – 2, neutral – 3, strong barrier – 4, very strong barrier – 5). The background study revealed that the generality of the sampled respondents possesses at least a master's degree qualification which connotes they are educationally qualified.

Data analysis

This study employed the factor analysis technique to determine variables that could be gauging the aspects of similar underlying dimensions and to identify clusters of related variables, thereby reducing them to a smaller logical framework (Norusis 2000; Ahadzie, Proverbs, and Olomolaiye 2008). Theoretically, factor analysis as a statistical technique can be utilized to analyze interrelationships (correlations) among many variables (e.g. questionnaire responses, test items, test scores) and to explain these variables in lieu of their related latent dimensions (factors). It is also described as a data reduction technique as it takes an extensive set of variables and seeks to summarize the data using a smaller set of factors/components (Pallant 2007). The study adopted exploratory factor analysis (EFA) as it aims to show any latent variables that can make the manifest variables co-vary. The shared variance of a variable is

Table 1: Definition of potential barriers.

Variable	Variable name
B1	Lack of well-defined biomimicry approach
B2	Lack of professional knowledge in biomimicry
B3	Lack of biomimicry in the university curriculum
B4	Lack of biomimicry database and information
B5	Lack of biomimicry training and education
B6	Lack of building codes and regulations
B7	Lack of government support
B8	Lack of incentives for biomimicry adoption
B9	Lack of biomimicry awareness
B10	Resistance to change in adopting new practices
B11	Lack of biomimicry technology
B12	Lack of multidisciplinary collaboration
B13	Perceived high cost of adopting biomimicry
B14	Lack of client demand
B15	Lack of biomimicry labelling/measurement framework
B16	Uncertainty on performance, efficiency and effectiveness of biomimicry
B17	Unavailability of biomimetic materials
B18	Potential time factor involved in implementation biomimicry
B19	Risk associated with implementation of biomimicry

Table 2: Communalities.

Stages	Initial	Extraction
Lack of well-defined biomimicry approach	1.000	.537
Lack of professional knowledge in biomimicry	1.000	.745
Lack of biomimicry in the university curriculum	1.000	.856
Lack of biomimicry database and information	1.000	.667
Lack of biomimicry training and education	1.000	.872
Lack of building codes and regulations	1.000	.747
Lack of government support	1.000	.660
Lack of incentives for biomimicry adoption	1.000	.503
Lack of biomimicry awareness	1.000	.659
Lack of biomimicry technology	1.000	.987
Perceived high cost of adopting biomimicry	1.000	.694
Lack of client demand	1.000	.570
Lack of biomimicry labelling/measurement framework	1.000	.850
Uncertainty on performance, efficiency and effectiveness of biomimicry	1.000	.718
Unavailability of biomimetic materials	1.000	.537
Potential time factor involved in implementation biomimicry	1.000	.729
Risk associated with implementation of biomimicry	1.000	.859

severed from its distinctive and error variance during factor extraction to show the underlying factor framework, thereby revealing just the shared variance in the result.

Factor analysis

Tables 2–6 and Figure 1 present the results from the EFA on the barriers hindering the adoption and implementation of biomimicry in the SACI. Out of the nineteen (19) variables listed, the following two (2) were omitted: resistance to change in adopting new practices (B.10) and lack of multi-disciplinary collaboration (B.12). In total, seventeen (17) variables were identified as potential barriers. The average communality of the variables in the study was above 0.6 (Table 2). The Kaizer-Meyer-Olkin (KMO) measure of sampling adequacy also attained a commendatory value of 0.76 (Table 3) while the Bartlett test of sphericity was significant, indicating that the population matrix was not an identity matrix (Table 4). The Cronbach's alpha value of 0.936 indicated the appropriateness of the research instrument used.

The data were subjected to PCA (with varimax rotation). The eigenvalue and factor loading were set at standard high values of 1.0 and 0.5, respectively. As shown in Table 5, four factors with eigenvalues exceeding 1.0 were extracted using the factor loading of 0.50 as the cut-off point. The factorability of the sample size for the analysis is supported by the relatively high values of the loading factor. The scree plot presented in Figure 1

Table 3: KMO and Bartlett's test.

Kaiser-Meyer-Olkin measure of sampling adequacy		.764
Emulate design principles	Approx. chi-square	1882.270
Brainstorm bio-inspired ideas	Df	136
Integrate life's principles	Sig.	.000

Table 4: Correlation matrix of factor analysis.

Factor	B1	B2	B3	B4	B5	B6	B7	B8	B9	B11	B13	B14	B15	B16	B17	B18	B19
B1	1.000																
B2	.386	1.000															
B3	.575	.333	1.000														
B4	.442	.459	.772	1.000													
B5	.370	.602	.525	.547	1.000												
B6	.637	.318	.386	.392	.48	1.000											
B7	.474	.518	.452	.656	.70	.515	1.000										
B8	.508	.383	.589	.565	.34	.257	.473	1.000									
B9	.309	.769	.365	.407	.76	.261	.497	.262	1.000								
B11	.612	.413	.861	.661	.24	.314	.306	.639	.286	1.000							
B13	.344	.556	.427	.595	.60	.444	.770	.361	.495	.395	1.000						
B14	.487	.423	.544	.427	.46	.442	.390	.501	.442	.624	.635	1.000					
B15	.503	.421	.386	.347	.51	.755	.427	.237	.324	.400	.612	.690	1.000				
B16	.490	.558	.478	.479	.55	.529	.470	.327	.374	.443	.642	.643	.766	1.000			
B17	.408	.493	.600	.522	.27	.239	.315	.502	.283	.701	.370	.469	.425	.490	1.000		
B18	.271	.603	.175	.297	.34	.468	.523	.138	.409	.284	.729	.378	.616	.584	.282	1.000	
B19	.430	.593	.285	.345	.32	.480	.425	.351	.373	.414	.636	.568	.681	.805	.357	.788	1.000

Note: Kaiser-Meyer-Olkin measure of sampling adequacy = .764; Bartlett test of sphericity = 1882.270; significance = 0.000. Explanation to acronyms as in Table 1.

Table 5: Rotated component matrix.[a]

	Component			
	1	2	3	4
Lack of biomimetic technology	.962			
Lack of biomimicry in the university curriculum	.842			
Limited availability of biomimetic materials	.658			
Lack of incentives for adopting biomimicry	.657			
Lack of database and information on biomimicry	.637			
Lack of well-defined biomimicry approach	.517			
Lack of client demand	.489			
Risk associated with implementation of biomimicry		.861		
Potential time factor involved in implementation of biomimicry		.778		
Uncertainty on performance, efficiency and effectiveness of biomimicry		.637		
Perceived high cost of adopting biomimicry		.538		
Lack of biomimicry training and education			.839	
Lack of biomimicry awareness			.739	
Lack of government support			.654	
Lack of professional knowledge in biomimicry			.606	
Lack of building codes and regulations				.770
Lack of biomimicry labelling/measurement framework				.646

Note: Extraction method: principal component analysis.
Rotation method: Varimax with Kaiser normalization.
[a]Rotation converged in 7 iterations.

also reveals the excluded factors by indicating the cut-off point at which the eigenvalues levelled off. The total variance explained by each of the extracted factors is as follows: Factor 1 (50.755%), Factor 2 (12.047%), Factor 3 (8.577%), and Factor 4 (6.278%) as shown in Table 6. Thus, the final statistics of the principal component analysis and the extracted factors accounted for approximately 72% of the total cumulative variance.

Discussion of results

Principal axis factoring revealed the presence of four (4) factors with eigenvalues above 1 as shown in Table 5. Based on the assessment of the innate relationships among the variables under each factor, the following interpretations were made. Factor 1 was termed *information and technology-related barriers*; Factor 2 *risk and cost-related barriers*; Factor 3 *knowledge-related barriers*; and Factor 4 *regulations-related barriers*. The names given these factors were derived from a close examination of the variables within each of the factors. The constituent indicators of each of the four factors extracted are explained below, together with a comprehensive description of each of them.

Factor 1: Information and technology-related barrier

As presented in Table 5, the seven (7) extracted barriers for Factor 1 were *lack of biomimetic technology* (96.2%), *lack of biomimicry in the university curriculum* (84.2%), *limited availability of biomimetic materials* (65.8%), *lack of incentives for adopting biomimicry* (65.7%), *lack of database and information on biomimicry* (63.7%), *lack of well-defined approach/strategy* (51.7%), and *lack of client demand* (48.9%). The number in parenthesis indicates the respective factor loadings. This cluster accounted for 50.8% of the variance. As noted by Darko and Chan (2016), lack of education, information, awareness, research, expertise and knowledge, as well as technological difficulties are top barriers to sustainable construction,

which aligns with the result of this study. Support for biomimicry discussions, workshops, research, seminars and training by stakeholders, driven through the various professional bodies and institutions, offers the potential to widely disseminate the requisite information on the importance of embracing this sustainability concept. It is inferred that improved financial support for research and development into eco-friendly products and technologies will help reduce and discourage the use of traditional ones.

Factor 2: Risk and cost-related barrier

The four (4) extracted barriers for Factor 2 were *risk associated with implementation of new practices* (86.1%), *potential time factor involved in implementation of new practices* (77.8%), *uncertainty on performance, efficiency and effectiveness of biomimicry* (63.7%), and *perceived high cost of adopting biomimicry* (53.8%) as presented in Table 5. The number in parenthesis indicates the respective factor loadings. This cluster accounted for 12.0% of the variance. The result here is in tandem with the study of Ametepey, Aigbavboa, and Ansah (2015) which listed fear of higher cost of investment, client worries about profitability, fear of a prolonged repayment period, lack of life-cycle cost knowledge, and lack of financial resources as the top barriers to the adoption and application of sustainable construction. There are many misconceptions about the subject of sustainability in the CI, a significant example of which is the myth that it is risky to undertake and that it costs more than conventional methods (Kubba 2012). However, a reality check on this myth revealed that in the long term, employing biomimicry to achieve sustainability is cost effective and affordable. This is authenticated by its considerations for waste reduction, energy saving, low carbon footprint and less pollution.

Factor 3: Knowledge-related barrier

This cluster accounted for 8.6% of the variance. The four (4) extracted barriers for Factor 3 were *lack of biomimicry*

Table 6: Component transformation matrix.

Factors	Initial eigenvalues			Extraction sums of squared loadings			Rotated sums of squared loadings		
	Total	% of variance	Cumulative %	Total	% of variance	Cumulative %	Total	% of variance	Cumulative %
1	**8.632**	50.775	50.775	8.358	49.163	49.163	3.927	23.099	23.099
2	**2.048**	12.047	62.822	1.815	10.674	59.837	3.307	19.453	42.553
3	**1.458**	8.577	71.399	1.214	7.139	66.976	2.983	17.549	60.102
4	**1.067**	6.278	**77.677**	.804	4.732	71.708	1.973	11.607	**71.708**
5	.806	4.743	82.420						
6	.673	3.957	86.377						
7	.580	3.411	89.789						
8	.435	2.561	92.350						
9	.397	2.338	94.688						
10	.290	1.704	96.392						
11	.203	1.192	97.584						
12	.123	.726	98.310						
13	.100	.586	98.895						
14	.079	.466	99.361						
15	.053	.312	99.673						
16	.033	.195	99.868						
17	.023	.132	100.000						

Note: Extraction method: principal axis factoring.

training and education (83.9%), *lack of biomimicry awareness* (73.9%), *lack of government support* (65.4%), and *lack of professional knowledge in biomimicry* (60.6%) as presented in Table 5. The number in parenthesis indicates the respective factor loadings. As affirmed by Pietrosemoli and Monroy (2013), sustainable construction is a result of collective attempts of construction professionals, investors, industry suppliers, and other stakeholders directed to the development of a built environment that considers environmental, socio-economic, energy and cultural conditions needed to bring integrated solutions to human society. However, the generality of these professionals and stakeholders lacks the requisite knowledge and training needed to implement sustainable construction practices. Also, it is discovered that there is no definite and conscious effort on their part to keep up-to-date with the continuously changing approaches to the construction process and activities that are environmentally responsive.

Factor 4: Regulations-related barrier
As presented in Table 5, only two (2) barriers were extracted for Factor 4, namely *lack of building codes and regulations* (77.0%), and *lack of labelling/measurement framework* (64.6%). The number in parenthesis indicates the factor loadings. This cluster accounted for 6.3%

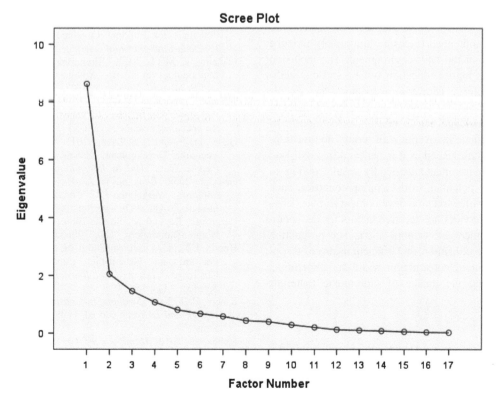

Figure 1. Scree plot for factor analysis.

of the variance. Codes, regulations, guidelines, and standards ensure a measurement critical for evaluating sustainability. In the absence of these, departures from the path and precepts of sustainable construction is bound to occur, thus hampering every effort towards achieving sustainability. Set rules and regulations, and green design guidelines and construction standards are identified by the study of AlSanad (2015) as imperative drivers in ensuring the adoption and application of sustainable construction practices, of which biomimicry has shown great potential.

Conclusion

Biomimicry proponents believe that the field of biomimicry will progress in a notable manner and eventually be universally embraced, owing to its vast domain of possible applications in different fields of human life (robotics, medicine, pharmaceuticals, engineering, and architecture). However, this paper intends to create a global biomimicry awareness in the CI where the application is at its infancy. As one of the sectors responsible for environmental degradation and pollution, the field of biomimicry has the potential to reposition and redirect the CI towards the green paradigm.

This paper presents the potential barriers to the adoption and application of biomimicry. The results of the analyzed data gathered reveal four main barriers. These are: *information and technology-related, risk and cost-related, knowledge-related, and regulations-related barriers*. Addressing these barriers head-on will help in maximizing the potential biomimicry has for enhancing the performance of buildings (smart buildings), maintaining biodiversity and ecosystem services, climate change adaptation and mitigation of greenhouse gas emissions.

Recommendation

Multidisciplinary collaboration coupled with stakeholder support are recommended to ensure biomimicry practice is well received in the built environment. The inclusion of biomimicry in higher education syllabi, and the facilitation of biomimicry education workshops and events are highly recommended for forging links and transfer of biomimicry knowledge across disciplines. Adoption of available biomimetic innovations, materials, and technologies is also recommended as it is believed this will lead to important environmentally-friendly results owing to their sustainable potential. In developing countries, such as South Africa, infrastructural development and the delivery of primary services are decisive factors for the socio-economic upliftment of its population. The production process, resource utilization, and efficient means of disposal and treatment of construction waste are imperative, thus necessitating the uptake of biomimetic materials and technologies.

ORCID

Olusegun A. Oguntona http://orcid.org/0000-0001-8963-8796

Clinton O. Aigbavboa http://orcid.org/0000-0003-2866-3706

References

Ahadzie, D., D. Proverbs, and P. Olomolaiye. 2008. "Critical Success Criteria for Mass House Building Projects in Developing Countries." *International Journal of Project Management* 26 (6): 675–687.

AlSanad, S. 2015. "Awareness, Drivers, Actions, and Barriers of Sustainable Construction in Kuwait." *Procedia Engineering* 118: 969–983.

Ametepey, O., C. Aigbavboa, and K. Ansah. 2015. "Barriers to Successful Implementation of Sustainable Construction in the Ghanaian Construction Industry." *Procedia Manufacturing* 3: 1682–1689.

Arnarson, P. O. 2011. *Biomimicry: New Technology*. Reykjavík University. Accessed November 16, 2016. http://olafurandri.com/nyti/papers2011/Biomimicry%20-%20P%C3%A9tur%20%C3%96rn%20Arnarson.pdf.

Azis, A. A. A., A. H. Memon, I. A. Rahman, S. Nagapan, and Q. B. A. I. Latif. 2012. "Challenges Faced by Construction Industry in Accomplishing Sustainability Goals." In *Business, Engineering and Industrial Applications (ISBEIA),2012 IEEE Symposium*, 630–634.

Aziz, M. S., and A. Y. El Sherif. 2015. "Biomimicry as an Approach for Bio-Inspired Structure with the aid of Computation." *Alexandria Engineering Journal* 55: 707–714.

Badarnah, L., and U. Kadri. 2015. "A Methodology for the Generation of Biomimetic Design Concepts." *Architectural Science Review* 58 (2): 120–133.

Benyus, J. M. 1997. *Biomimicry: Innovation Inspired by Nature*. New York, USA: William Morrow & Company.

Benyus, J. M. 2011. *A Biomimicry Primer*. Missoula, Montana: The Biomimicry Institute and the Biomimicry Guild.

Darko, A., and A. P. Chan. 2016. "Review of Barriers to Green Building Adoption." In *Sustainable Development*. John Wiley & Sons, Ltd and ERP Environment.

De Pauw, I., P. Kandachar, E. Karana, D. Peck, and R. Wever. 2010. *"Nature Inspired Design: Strategies Towards Sustainability."* Knowledge collaboration & learning for sustainable innovation: 14th european roundtable on sustainable consumption and production (ERSCP) conference and the 6th environmental management for sustainable universities (EMSU) conference, Delft, The Netherlands, October 25–29. Delft University of Technology; The Hague University of Applied Sciences; TNO.

El Din, N. N., A. Abdou, and I. A. El Gawad. 2016. "Biomimetic Potentials for Building Envelope Adaptation in Egypt." *Procedia Environmental Sciences* 34: 375–386.

El-Zeiny, R. M. A. 2012. "Biomimicry as a Problem Solving Methodology in Interior Architecture." *Procedia - Social and Behavioral Sciences* 50: 502–512.

Gamage, A., and R. Hyde. 2012. "A Model Based on Biomimicry to Enhance Ecologically Sustainable Design." *Architectural Science Review* 55 (3): 224–235.

Giang, D. T., and L. S. Pheng. 2011. "Role of Construction in Economic Development: Review of Key Concepts in the Past 40 Years." *Habitat International* 35 (1): 118–125.

Goss, J. 2009. *Biomimicry: Looking to Nature for Design Solutions*. Washington DC: Corcoran College of Art and Design, ProQuest Dissertations Publishing.

Hargroves, K., and M. Smith. 2006. "Innovation Inspired by Nature: Biomimicry." *Ecos* 2006 (129): 27–29.

Hussin, J. M., I. A. Rahman, and A. H. Memon. 2013. "The Way Forward in Sustainable Construction: Issues and Challenges." *International Journal of Advances in Applied Sciences* 2 (1): 15–24.

Kibert, C. J. 2013. *Sustainable Construction: Green Building Design and Delivery*. 3rd ed. Hoboken, NJ: John Wiley & Sons.

Kubba, S. 2012. *Handbook of Green Building Design and Construction: LEED, BREEAM, and Green Globes*. Oxford: Butterworth-Heinemann.

Marshall, A. 2013. "Biomimicry." In *Encyclopedia of Corporate Social Responsibility*, 174–178. Berlin Heidelberg: Springer.

Minnesota Inventors Hall of Fame. 2016. *Inductees, Dr Otto H. Schmitt–1978 Inductee.* Accessed February 27, 2016. http://www.minnesotainventrs.org/inductees/otto-h-schmitt. html.

Murr, L. E. 2015. "Biomimetics and Biologically Inspired Materials." In *Handbook of Materials Structures, Properties, Processing and Performance*, New York City: 521–552. Springer International Publishing.

Norusis, M. J. 2000. *SPSS 10.0 Guide to Data Analysis.* New York: SPSS Inc.

Nychka, J. A., and P. Chen. 2012. "Nature as Inspiration in Materials Science and Engineering." *JOM Journal of the Minerals, Metals and Materials Society* 64 (4): 446–448.

Oguntona, O. A., and C. O. Aigbavboa. 2016. "Promoting Biomimetic Materials for a Sustainable Construction Industry." *Bioinspired, Biomimetic and Nanobiomaterials* 6 (3), 122–130.

Okuyucu, C. 2015. "Biomimicry Based on Material Science: The Inspiring art from Nature (Review Article)." *Matter* 2 (1): 49–53.

Pallant, J. 2007. *SPSS Survival Manual: A Step-by-Step Guide to Data Analysis Using SPSS Version 15.* Nova Iorque: McGraw Hill.

Pearce, A., and Y. H. Ahn. 2012. *Sustainable Buildings and Infrastructure: Paths to the Future.* New York: Routledge.

Pietrosemoli, L., and C. R. Monroy. 2013. "The Impact of Sustainable Construction and Knowledge Management on Sustainability Goals. A Review of the Venezuelan Renewable Energy Sector." *Renewable and Sustainable Energy Reviews* 27: 683–691.

Plank, R. J. 2005. Sustainable Construction - A UK Perspective. *Structures Congress 2005 held on 20–24 April in New York, United States.* 1–7.

Pronk, A., M. Blacha, and A. Bots. 2008. "*Nature's Experiences for Building Technology.*" Proceedings of the 6th international seminar of the international association for shell and spatial structures (IASS) working group.

Ramsaran, R., and R. Hosein. 2006. "Growth, Employment and the Construction Industry in Trinidad and Tobago." *Construction Management and Economics* 24 (5): 465–474.

Rinaldi, A. 2007. "Naturally Better. Science and Technology are Looking to Nature's Successful Designs for Inspiration." *EMBO Reports* 8 (11): 995–999.

Shi, L., K. Ye, W. Lu, and X. Hu. 2014. "Improving the Competence of Construction Management Consultants to Underpin Sustainable Construction in China." *Habitat International* 41: 236–242.

Shi, Q., J. Zuo, and G. Zillante. 2012. "Exploring the Management of Sustainable Construction at the Programme Level: A Chinese Case Study." *Construction Management and Economics* 30 (6): 425–440.

Shu, L., K. Ueda, I. Chiu, and H. Cheong. 2011. "Biologically Inspired Design." *CIRP Annals – Manufacturing Technology* 60 (2): 673–693.

Singh, A., and N. Nayyar. 2015. "Biomimicry-an Alternative Solution to Sustainable Buildings." *Journal of Civil and Environmental Technology* 2 (14): 96–101.

Strategic Direction. 2008. "Nature's Inspiration: Solving Sustainability Challenges." *Strategic Direction* 24 (9): 33–35.

Tam, C., V. W. Tam, and W. Tsui. 2004. "Green Construction Assessment for Environmental Management in the Construction Industry of Hong Kong." *International Journal of Project Management* 22 (7): 563–571.

Vincent, J. F., O. A. Bogatyreva, N. R. Bogatyrev, A. Bowyer, and A. K. Pahl. 2006. "Biomimetics: Its Practice and Theory." *Journal of the Royal Society, Interface / The Royal Society* 3 (9): 471–482.

Wang, N. 2014. "The Role of the Construction Industry in China's Sustainable Urban Development." *Habitat International* 44: 442–450.

Zari, M. P. 2007. *Biomimetic Approaches to Architectural Design for Increased Sustainability.* SB07 Auckland, New Zealand.

Zari, M. P. 2008. "*Bioinspired Architectural Design to Adapt to Climate Change.*" Proceedings of the world conference. SB08 (ISBN 978-0-646-50372-1): 771–8.

Inventory of kiln stacks emissions and health risk assessment: Case of a cement industry in Southwest Nigeria

Solomon O. Giwa ⓘ , Collins N. Nwaokocha ⓘ and Abayomi T. Layeni

Cement production is a significant source of air pollution as both gaseous and particulate materials released are detrimental to the ecosystem. This work was carried out in a cement industry located in Southwest Nigeria. The emission rates of carbon monoxide (CO), nitrogen oxides (NO_x), carbon dioxide (CO_2) and sulphur oxides (SO_x) released from the cement kilns using fuel oil, natural gas (NG) and coal were garnered for a year. Thereafter, the estimated emission quantities of the pollutants were employed to obtain the emission inventory of the cement plant. Uncertainty analysis associated with the emissions was evaluated using Analytica® (4.6). Total amounts of pollutants emitted from the plant were 4.86 tonne (t) (NO_x), 18.2 t (SO_x), 2.270 Kt (CO_2) and 1.17 t (CO). Uncertainty range of −149.38% to 149.38% was connected to all the pollutants. Results showed that the quantities of pollutants discharged from the cement industry were considerably higher than recommended. The evaluated air quality indices for CO, NO_x, and SO_x implied that the health risk on exposure to these gases was hazardous. This study revealed that NG and wastes are the best fuel for kiln firing to reduce the amounts of pollutants emitted into the microenvironment of the plant.

Introduction

Fossil fuels of coal, natural gas (NG), petroleum coke, and fuel oil (FO) are the main global energy sources used in the industrial sector, especially cement manufacturing industries (Uson et al. 2013; Chatziaras, Psomopoulos, and Themelis 2016). A cement kiln consumes around 30–40% of the total energy utilized for cement production. The choice of a fuel largely depends on the country, environmental effect, storage, processing, handling, fuel type, cost, and availability. Cement plants typically are renowned for their intensive raw material, fuel (fossil), carbon, energy consumption, and emission (Pudasainee et al. 2009; Uson et al. 2013; Barcelo et al. 2014). It has been reported that the global cement sub-sector accounts for about 10–15% of the energy used in the industrial sector globally, reaching a peak of 120 kWh/t of cement (Uson et al. 2013). Subject to the fuel type, 60 kg–150 kg of the fuel is consumed to produce a tonne of cement (Cembu and theoretically requiring 1750 MJ to produce a tonne of clinker (Hendriks et al. 2002). Over 150 countries of the world manufacture cement and/or clinker, with China ranked first (661 MMt), followed by India (100 MMt) and the United States (90 MMt) in 2001 (Shen et al. 2014).

Emissions of gases (carbon dioxide (CO_2), carbon monoxide (CO), sulphur oxides (SO_x), nitrogen oxides (NO_x), methane (CH_4), hydrogen sulphide (H_2S) etc.), particulates (black carbon, organic carbon, benzene, particulate matters etc.), dusts and heavy metals (mercury, arsenic, chromium, lead etc.; when wastes are used) from cement industries are of serious global concern as these pollutants are the major cause of environmental pollution. Sources of emissions in the industry include: excavation works, dumping and tipping activities, conveyor belts, kilns, raw material and cement mills, of which kiln emissions (process-related and fuel combustion) is mainly responsible for the emission of pollutants.

Globally, cement production is one of the highest emitters of CO_2 (Benhelal et al. 2013). The production of cement accounted for the discharge of 2.37 Gt of pollutants into the atmosphere in the year 2000. About 1.8 Gt of CO_2 (6%) out of around 28.3 Gt of global CO_2 estimate was emitted in 2015 (Benhelal et al. 2013; Uson et al. 2013). However, this value was reduced to 5% in the last few years due to the utilization of alternative materials and fuels, and energy efficiency improvements (Chatziaras, Psomopoulos, and Themelis 2016).

Numerous studies have been conducted on cement production relating to energy efficiency and savings, reduction of pollutants' emission, inventory and assessment of emissions, usage of alternative materials and fuels, modelling of energy recovery and emission forecasting, environmental impact assessment and life cycle of greenhouse gas (GHG) emission (Pudasainee et al. 2009; Chen et al. 2010; Atabi, Ahadi, and Bahramian 2011; Lei et al. 2011; Ilalokhoin et al. 2013; Uson et al. 2013; García-Segura, Yepes, and Alcalá 2014; Shen et al. 2014; Abdul-Wahab et al. 2016; Chatziaras, Psomopoulos, and Themelis 2016). Atabi, Ahadi, and Bahramian (2011) employed scenario analysis to examine the effect of different policies in reducing CO_2 released from the cement industry in Iran. They found that an integrated scenario resulted in the highest CO_2 reduction (13%) as against the business as usual scenario compared to those of fuel switching (4.9% reduction) and energy efficiency (9.8%) for a period of 15 years (2005–2020). A study on the emission of harmful air pollutants (volatile organic compounds and heavy metals) from three cement kilns fuelled by co-burning waste showed that the pollutants were within the recommended emission limits with efforts towards their reduction being made (Pudasainee et al. 2009).

In addition, the release of CO_2 from a cement plant was modelled and used to assess its effect on the workplace and microenvironment (Abdul-Wahab et al. 2016). It was observed that the maximum level of CO_2 predicted by the model for a period of 1 hr was higher than the acceptable levels for the selected days (winter and summer) while the CO_2 emissions within and outside the plant were found to be higher for the line sources than the point sources during winter days. Shen et al. (2014) employed a factory-level sampling technique (Tier three) for the first time to obtain the CO_2 emission inventory of China using a bottom-up approach for various types of cement and clinker production. They reported that earlier studies overestimated carbon emissions from cement production in China because technology transition (wet process to dry process), substitution of raw materials and fuels, lime content disparity, blend additive utilization, and clinker-to-cement ratios were neglected in their works.

Studies on characterization, modelling and inventory of emissions from cement plants in Nigeria are very scarce in the public domain despite the global concern raised over the enormous and adverse effect of pollutants related to this source as mostly reported in the literature (Pudasainee et al. 2009; Chen et al. 2010; Lei et al. 2011; Chatziaras, Psomopoulos, and Themelis 2016). Amos et al. (2015) and Oyinloye (2015) both examined the effects of cement dust and emissions on the soil properties (physicochemical) and inhabitants within 10 km radius of Ashaka Cement Company Plc., located at Ashaka, Gombe State (Nigeria) and Lafarge West African Portland Cement Company (WAPCO) (Ewekoro plant) in Ogun State, Nigeria, respectively. Both studies reported the serious impact of the cement particulate on the soil, environment and health risk of inhabitants within a 5 km radius of both plants. Ideriah and Stanley (2008) experimentally evaluated the concentrations of suspended particulate matter (SPM) and NO_2 in the air around Atlas and Eagles cement industries located at Port Harcourt, Nigeria. Their results revealed that the concentrations of SPM and NO_2 at Atlas cement were considerably lower than those at Eagle cement, with SPM concentrations in both industries and NO_2 at Eagle cement higher than the recommended limits. A similar study was conducted by Dada, Olatunde, and Oluwajana (2013) to assess the air quality around Ewekoro cement plant of WAPCO based in Ogun State, Nigeria at various distances (0–1500 m) from the plant location. They observed significant levels of SPM, particulates ($PM_{2.5}$ and PM_{10}), SO_x, NO_x, CO, and H_2S, which decreased with an increase in distance from the plant.

Previous studies have shown that the release of pollutants generated due to activities of the cement industries locally and globally are sources of serious environmental pollution (Lei et al. 2011; Benhelal et al. 2013; Shen et al. 2014; Abdul-Wahab et al. 2016). Air pollution from cement industries has been a prominent source of global warming and climate change (Uson et al. 2013). Globally, efforts have been geared towards the reduction of the quantities of pollutants discharged into the atmosphere via cement industries. Notable emission mitigation strategies include improvements of energy efficiency, switching from conventional fuels to alternative fuel (waste), emission capture and storage, raw material and cement blending with other materials (Pudasainee et al. 2009; Lei et al. 2011; Chatziaras, Psomopoulos, and Themelis 2016;). These schemes are primarily accompanied with cost, energy and environmental pollution reduction (Chatziaras, Psomopoulos, and Themelis 2016). Of these mitigation strategies, the substitution of traditional fuels with alternative fuels has been noted to be a key emission reduction technique. The use of wastes as fuel in cement kilns in the USA, Australia, Canada, Korea, Japan, and European countries (Ali, Saidur R, and Hossain 2011) is an established technology which is yet to be replicated in Nigeria and other African countries. The various wastes generated in Nigeria because of the country's large human population and the lack of facilities to recycle these huge wastes call for urgency in employing this potential and untapped energy source as a fuel for cement production in the country.

However, there is a scarcity of documentation on the epidemiological study of the effect of emissions on public health of residents near cement plants in the country. The data from the national emission inventory for the year 2000 revealed that cement industries are the major contributors of CO_2 emissions in the industrial processes sub-sector (64.52%), which signalled the urgent need to curb emissions from this source (Nigeria's Second National Communication 2014). Due to the alarming growth and tonnage of cement industries in Nigeria, and the utilization of traditional fuels which further portends an increase in environmental pollution and degradation subject to the discharge of obnoxious gases and particulates into the microenvironment, there arises the necessity to undertake an emission inventory, and to assess stack emissions and health risks due to cement production activities. In this present study, emission inventory, cement stack emission characterization, and associated health risk were undertaken for a major cement plant in Southwest Nigeria using three fuels for kiln firing.

Background of cement production in Nigeria

The production of cement in Nigeria dates back to 1957 with three plants being commissioned at different times by the Mid-Western (WAPCO), Eastern (Benue Cement Company (BCC)), and Northern (Cement Company of Northern Nigeria (CCNN)) regional governments. Cement production in Nigeria commenced as a way of substituting cement importation because of the huge domestic demand for cement. Presently, there is a progressive increase in the number of cement industries in Nigeria, which are distributed along the limestone deposit belts of the country. Key players in the industry include; Lafarge WAPCO Nigeria Plc, Ashaka Cement Company Plc, Edo Cement Company Limited, BCC, CCNN, Calabar Cement Company Limited, Nigerian Cement Company Limited, and Dangote Cement Limited. Cement demand in Nigeria has been reported to be the largest in sub-Saharan Africa with nearly 95% of the raw materials utilized to produce cement available locally (Industry Report 2011).

The energy (electricity) problem bedeviling the country is taking a critical toll on the cement industries in the country. This has led these industries like others to generate their own power as this sub-sector of the economy is energy intensive. Coal, NG and FO are being utilized as alternative energy sources to the public electricity for cement production. The amounts and types of pollutants released into the environment of cement plants are strongly linked to the composition (carbon and elemental components) of the fuel and raw materials, combustion conditions and variables, environmental and meteorological factors, and other relevant parameters. The challenges experienced by the cement manufacturers are an upsurge in the prices of fuel and scarcities in supply. This has resulted in the dwindling of both profits and production outputs of most of the plants. The cement companies reacted to these challenges differently. For example, AshakaCem switched to the use of coal by engaging in a coal-mining venture due to the high cost and unstable supply of low pour fuel oil (LPFO). Doing this, the overall energy cost was reduced, and the fuel supply was guaranteed to a large extent. The refurbishment of the roller press was also carried out to enhance the cement mill efficiency and reduce power consumption during cement grinding. In a similar reaction, Lafarge WAPCO switched to the use of NG to fire their kilns and generate power in order to guarantee the supply of power and hence, cut energy costs.

Materials and methods

Area of study and plant operation

This work was conducted at a cement industry sited in Southwest Nigeria, which is one of the major players and brands in the country. The sole purpose of non-declaration of the cement industry's name is to protect it against any form of issue, especially environmental matters that may arise from the outcome of this work. This present study was carried out between January and October 2016, which translated to a year as the two remaining months were used for annual general maintenance of the cement plants. The operating hours for the year under consideration was 7320 h (305 days). The cement factory has two plants (three production lines) with a combined capacity of over 2 million tonnes of cement annually. The kiln in each production line of the cement industry is fired using coal, FO, and NG as fuels.

Data source and collection

Data on cement kiln stack emissions spanning a decade were sought from many of the cement industries in the country. All the contacted industries turned down our requests citing company's policies and implications of such data as reasons. However, our untiring efforts yielded the data utilized in this study with the condition of anonymity regarding the specific source of data. Daily records (emission rates) of the point source emissions of a GHG (CO_2) and precursor gases (CO, SO_x, and NO_x) emanating from the kiln end of each production line of the cement industry through the stack were collected. The obtained emission rates were for a period of 7320 h of running of the plants. The kiln stack emissions were measured at a time-weighted average concentration using the various gas analyzer sensors embedded into the cement kiln system. The stack emission rates were computed automatically with the use of other parameters such as fuel consumption, emission factor, clinker tonnage, emission concentrations, etc. as described in the literature (Abdul-Wahab et al. 2016). In addition, the data on total cement produced per annum spanning 26 years (1986 to 2011) and CO_2 emitted (1958 to 2014) from cement production for the country were sourced from the literature (Mojekwu, Idowu, and Sode 2013; US NOAA). The quantity and concentration (Srujan 2014) of each pollutant released into the atmosphere due to the fuels used in the kilns for clinker production were calculated as expressed in Eqs. (1) and (2), respectively, from the obtained emission rates. The total amount of all the pollutants released from the kiln stacks was also estimated using Eq. (3).

$$E_{Q_{i,j}} = E_{R_{i,j}} \times t \qquad (1)$$

$$P_{C_{i,j}} = 1.81 E_{R_{i,j}} \qquad (2)$$

$$E_T = \sum_{i=1,j=1}^{3} E_{Q_{i,j}} \qquad (3)$$

where:

i, j = fuel (NG, FO, and coal) and pollutant (NO_x, CO, SO_x, and CO_2), respectively;

$E_{R_{i,j}}$ Emission rate for individual fuel and pollutant (kg/day or kg/month);

$E_{Q_{i,j}}$ Emission quantity for individual fuel and pollutant (kg);

$P_{C_{i,j}}$ Concentration of individual pollutant and fuel (mg/m^3);

E_T Total emission quantity from the cement plant (kg or tonne);

t = time (h).

Data analysis

The daily and monthly averages and the total of the emission rates, quantities and concentrations for the pollutants (CO_2, CO, SO_x, and NO_x) associated with the three fuels used in kiln firing were analyzed. These were carried out for both the fuels and pollutants. Statistical analysis involving analysis of variance (ANOVA) and correlations were performed on the emission rates, quantities and concentration data obtained for all the pollutants from the fuels utilized in the cement kilns and other data (CO_2 emissions and cement production) used in this study.

Procedure for emission uncertainty evaluation

The uncertainty associated with the release of the kiln stacks emissions (CO_2, CO, SO_x, and NO_x) in the studied cement industry was estimated using the emission (quantity) data as the input variables. The emission data for each individual pollutant is the sum of the emissions from all the cement kilns (fuelled using NG, FO, and coal) for the period of operation of the kilns (7320 h).

The fitting of the emission data (input variables) into the suitable probability distribution function and the modelling of the emission uncertainty estimates were carried out using EasyFit® 5.6 (evaluation version) and Analytica® (4.6) software, respectively. Emission data of each pollutant were fed into EasyFit® 5.6 to perform the probability distribution fitting. The best fitted distribution as ranked by the software and the corresponding values were thereafter input into the Analytica® for uncertainty estimation modelling. The use of Analytica® consisted of the input, uncertainty propagation (input variables) and output stages. The input model was developed using the fitted probability distribution of the input variables. Tier 2 method was employed as it has been endorsed for national GHG inventories (IPCC 2006) reporting and computation. Hence, the choice of Analytica® software and the use of Latin hypercube sampling (LHS) for simulation of the input model. In this study, LHS was selected over Monte Carlo Simulation (MCS) as a numerical simulation technique because it offers a better approximation with large number of samples. The propagation of the input model consisting of the probability distributions of the input variables was performed using LHS. Median

Latin hypercube and Minimal Standard were set as the default for sampling and random number generation, respectively. Iterations of the simulation was carried out until there was no further change in the value of the standard deviation. Thereafter, the mean value of each output variable was obtained from the simulation. Further information for estimating emission uncertainty can be read in previous studies published by Giwa, Nwaokocha, and Layeni (2016; Forthcoming). The uncertainty related to the release of the pollutants was estimated using Eq. (4). A flow chart of this study including the modelling procedure is provided in Figure 1.

$$R \ (\%) = \left(\frac{X - S}{S}\right) \times 100 \ \ and \ \left(\frac{Y - S}{S}\right)$$
$$\times 100 \quad\quad (4)$$

where:

X = 2.5th percentile of mean (simulated)
Y = 97.5th percentile of mean (simulated)
S = Mean (simulated)
R = Relative uncertainty

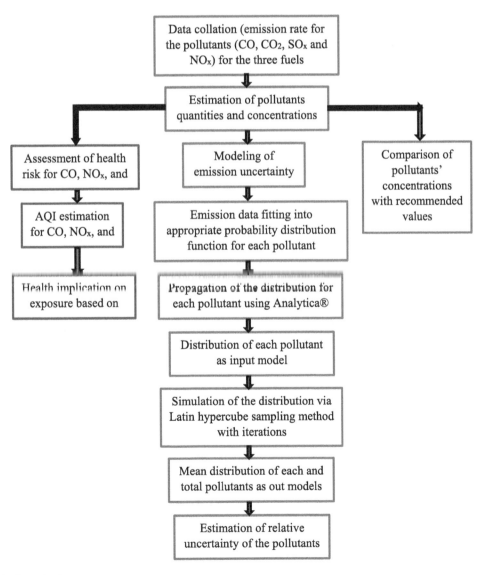

Figure. 1: Study flow chart.

Table 1: Air quality index descriptor.

Breakpoints					
$PM_{2.5}$ ($\mu g/m^3$)	CO (ppm)	SO_2 (ppm)	NO_2 (ppm)	AQI	Category
0.0–15.4	0.0–4.4	0.000–0.034	–	0–50	Good
15.5–40.4	4.5–9.4	0.035–0.144	–	51–100	Moderate
40.5–65.4	9.5–12.4	0.145–0.224	–	101–150	Unhealthy
65.5–150.4	12.5–15.4	0.225–0.304	–	151–200	Unhealthy
150.5–250.4	15.5–30.4	0.305–0.604	0.65–1.24	201–300	Very unhealthy
250.5–350.4	30.5–40.4	0.605–0.804	1.25–1.64	301–400	Hazardous
350.5–500.4	40.5–50.4	0.805–1.004	1.65–2.01	401–500	Hazardous

Source: Rim-Rukeh, 2015

Air quality index estimation

A comparison of the obtained results for this study with outdoor air quality standards (American Society of Heating, Refrigerating, and Air-Conditioning Engineers (ASHRAE) and World Health Organization (WHO)) was carried out. This was to examine their compliance with the stipulated values prescribed by these standards. In addition, an assessment of the health risk connected to NO_x, SO_x, and CO was carried out with the use of Eq. (5) as given in the literature for estimating air quality index (AQI) (Rim-Rukeh, 2015). Table 1 presents the AQI signifier.

$$I_p = (C_p - BP_L)\left(\frac{I_H - I_L}{BP_H - BP_L}\right) + I_L \quad (5)$$

where:

I_P = pollutant index;
C_P = rounded pollutant concentration;
BP_H = breakpoint equal or greater than C_P;
BP_L = breakpoint equal or less than C_P;
I_H = AQI value matching BP_H;
I_L = AQI value matching BP_L;

Results and discussion
Kiln stack emissions from clinker production

The environmental issue is one of the major problems facing cement production globally. Several pollutants are released through cement kiln stacks into the atmosphere which impacts the environment adversely and causes global warming, climate change, acid rain, and respiratory and heart-related health diseases, deterioration of air quality, etc. Data on emission rates of CO_2, CO, SO_x, and NO_x released from the cement plants under consideration were obtained via clinker formation in the kilns operated using three fuels (coal, FO (low pour) and NG). It is worth noting that CO and NO_x as reported in this work, were direct products of fuel combustion while the CO_2 and SO_x were sourced from both combustion of the fuels, and calcination of CO_2 and SO_x reaction with limestone during clinker formation in the cement kilns.

NO_x emission rates

The sum of the daily emission quantity of NO_x released via kiln stack due to clinker production for each month of the 7320 h operation of the cement kilns is presented in Figure 2. The daily mean emission of NO_x released from the kilns was estimated to be 1.62 kg (2.94 mg/m³) (NG), 7.08 kg (12.82 mg/m³) (FO) and 7.23 kg (13.08 mg/m³) (coal) for the period of operation of the plants. In this study, 4.86×10^3 kg of NO_x (consisting of 10.18% NG, 44.45% FO, and 45.36% coal) was released from the plants for 7320 h of operation. The highest amounts of NO_x for coal, FO and NG were recorded in the months of October (232.7 kg), August (227.8 kg), and January (57.3 kg), respectively. Similarly, the lowest quantities of NO_x emitted from the plants were in the months of February, January, and April for coal, FO, and NG, respectively (see Figure 2). The total NO_x released in relation to the fuels was 495.3 kg, 2160.77 kg, and 2205.2 kg for NG, FO, and coal, respectively. It can be seen from Figure 2 that coal and FO emitted significantly higher amounts of NO_x compared to NG as fuels used for firing the cement kilns. This can be connected to the gaseous nature and carbon content of NG. Due to stable ventilation and the high operating temperature of rotary kilns, a high quantity of NO_x has been reported regarding this kiln type (Lei et al. 2011). In addition, the obtained values of NO_x in this work are moderately lower than those of cement plants with similar output capacities reported in the literature, though the fuel type used, and other features of these plants were not given (Shen et al., 2005).

The concentrations of NO_x for all the fuels reported in this work are considerably higher than the daily maximum limit (50 µg/m³; daily) recommended by WHO (WHO 2000). Also, the IAQ index of the emission of NO_x from the cement industry to the atmosphere gave a value higher than 500, implying that the health risk linked to this pollutant on exposure is hazardous. NO_x is mainly formed by the oxidation of atmospheric nitrogen during fuel combustion and the chemical reaction of raw materials, mostly at a very high temperature in the presence of oxygen. Consequently, acid rain and smog, and the formation of secondary pollutants such as black carbon and ozone in the atmosphere around the cement industry, due to the excessive and long-time discharge of NO_x, are highly anticipated. This will negatively impact the residents, microenvironment and ecosystem causing chronic respiratory infection and pollution of air, soil, water, and vegetation in the area where the cement industry is located.

SO_x emission rates

The monthly emission amounts of SO_x from the fuels used in the production of clinker in the studied plants is

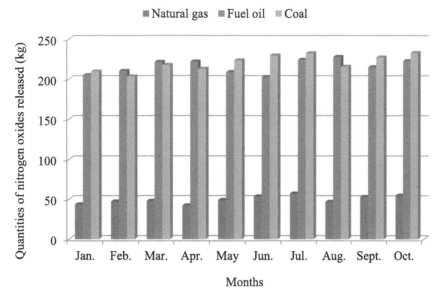

Figure 2: NO_X released into the atmosphere (kg/month).

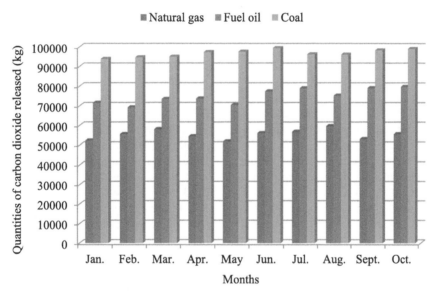

Figure 3: CO_2 released into the atmosphere (kg/month).

illustrated in Figure 3. The total and daily averages of SO_x released for the operation of the kilns was 5.82 kg and 0.019 kg (0.035 mg/m^3) for NG, 5685.27 kg and 18.64 kg (33.73 mg/m^3) for FO, and 12549.33 kg and 41.15 (74.45 mg/m^3) for coal, respectively. Estimated total emission of SO_x from the plants was 1.82×10^4 kg, consisting of 0.03% from NG, 31.17% from FO, and 68.80% from coal. For NG, FO and coal, maximum and minimum amounts of SO_x released monthly were 1.02 kg (September) and 0.07 kg (July), 643.6 kg (September) and 469.3 kg, and 1381.4 kg (April) and 1161.9 kg (June), respectively. SO_x emitted from the use of NG as fuel is very low and insignificant compared to those of coal and FO, with the quantities of SO_x released from coal more than twice that of FO for most of the months (Figure 3). This can be connected to the very low sulphur content of NG in comparison to moderate and high sulphur content of FO and coal, respectively. The obtained emission quantities for the period considered

in this study are moderately less than those reported in previous works with similar plants' tonnage (Shen et al., 2005). It is pertinent to know that the characteristics of these plants were not given to enhance better comparison.

With the estimated daily concentrations of SO_x for the fuels, it is observed that only the kiln fired using NG has SO_x values moderately less than the maximum (50 μg/m^3; daily) prescribed by WHO (WHO 2000). This shows that the quantities of SO_x discharged via the kiln stack into the atmosphere using FO and coal as fuels are detrimental to the environment. The sources of SO_x are from the burning of fuels and chemical reactions of raw materials in the kiln system. The emission of SO_x causes acid rain and secondary pollutants' formation, which adversely affects physical structures (building walls and roofs), ecosystems, the environment and human health (acute respiratory infection), is probably highly pronounced in the microenvironment where the industry is located. The estimation of the AQI, which

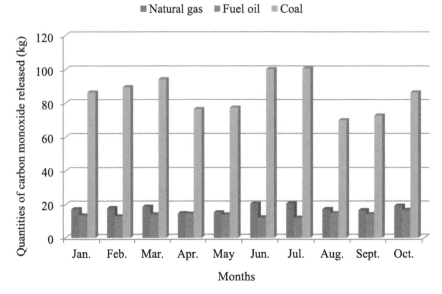

Figure 4: CO released into the atmosphere (kg/month).

indicates the health risk connected to this pollutant, was higher than 500, meaning, exposure to SO_x released from the cement industry is hazardous.

CO_2 emission rates
The calcination process and fossil fuel combustion are responsible for close to 50% and 40%, respectively, of the total CO_2 emissions during the process of cement manufacturing (Benhelal et al. 2013; Barcelo et al. 2014; Abdul-Wahab et al. 2016). Figure 4 presents the emission quantities of CO_2 released monthly by producing cement clinkers using coal, NG, and FO as fuels. When coal was used as fuel, an estimated total amount of CO_2 discharged from the kiln stack was 5.55×10^5 kg, with daily and monthly averages of 1.82×10^3 kg (3.30×10^3 mg/m³) and 5.55×10^4 kg, respectively. For FO and NG, however, the total, daily and monthly averages of CO_2 emissions were 7.50×10^5 kg, 2.46×10^3 kg (4.45×10^3 mg/m³) and 7.50×10^4 kg, and 9.67×10^5, 3.17×10^3 kg (5.74×10^3 mg/m³) and 9.67×10^4 kg, respectively. The estimated total CO_2 discharged from the plants was 2.27×10^6 kg (24.4%; NG, 33.0%; FO, 42.9%; coal). As can be noticed in Figure 4, CO_2 emitted from all the fuels are more (in magnitude) compared to those of SO_x and NO_x as presented in Figures. 2 and 3. The amounts of CO_2 discharged from the production of tonnes of clinker were highest when coal was used as fuel, followed by FO and then NG. This is directly linked to the carbon contents of the fuels (coal > FO > NG) as the same quality of limestone and other raw materials were used in clinker production.

With a daily outdoor CO_2 concentration of 1.35×10^4 mg/m³, which is more than the 100 mg/m³ (100,000 ppm) considered to be very poor air quality, it is therefore apparent that the quantities of CO_2 released via cement kiln production from the plants are enormous. This contributes considerably to the local, national and regional source of CO_2, which causes global warming leading to climate change. Furthermore, cement production has been recognized as one of the foremost sources of CO_2 emission which leads to both global warming and climate change (Uson et al. 2013).

CO emission rates
All forms of fuel combustion have to do with the emission of CO, whether complete or incomplete. The burning of NG, FO, and coal as fuels in the cement kilns is always associated with CO emissions. The emissions of CO on monthly basis from these fuels are illustrated in Figure 5. The estimated daily means of CO released from the plants were 0.588 kg (1.06 mg/m³), 0.460 kg (0.83 mg/m³), and 2.80 kg (5.07 mg/m³) for the use of NG, FO, and coal, respectively. The highest amounts of CO emitted were recorded in the months of July (20.79 kg) for NG, October (16.92 kg) for FO and July (100.76 kg) for coal whereas the lowest quantities were observed in the months of May (14.12 kg) for NG, July (12.16 kg) for FO and August (69.98 kg) for coal. In the context of this work, the amounts of CO released from the kilns were 179.22 kg (NG), 140.4 kg (FO) and 854.26 kg (coal) for the duration considered in operating the plants. Also, the total CO released from the plants was 1.17×10^3 kg (15.27%; NG, 11.95%; FO, and 72.78%; coal). Coal is observed to emit the highest amounts of CO in the order of four to five and four to eight compared with NG and FO, respectively (Figure 5). This is largely due to the high carbon content and solid nature of coal in comparison to those of NG and FO. The slight increase in the amount of CO released from the use of NG as fuel in clinker production relative to that of FO (see Figure 5) can be linked to the pre-treatment given to FO prior to its use for firing cement kiln. Compared to the other pollutants investigated in the work, CO is noticed to have the lowest magnitude of emission. This agrees with the literature stating that rotary kilns release lower quantities of CO compared to shaft kilns due to steady air inflow and increased operating temperatures (Lei et al. 2011).

The obtained daily concentration values of CO released from the plants are far higher than the

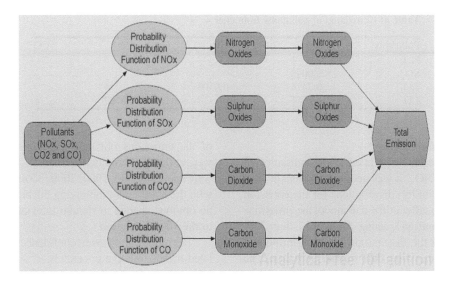

Figure 5: Influence diagram of developed model for uncertainty estimation.

Table 2: Correlation of pollutants from NG.

Pollutants	NO_x	SO_x	CO_2	CO
NO_x	1			
SO_x	0.0885	1		
CO_2	0.1539	−0.0500	1	
CO	0.5186	−0.4466	0.5804	1

recommendation (60 $\mu g/m^3$; daily maximum) of WHO (WHO, 2010). Therefore, the CO released from the plants into the environment portends danger to the micro-environment and public health (causing congestion of brain and lungs, tuberculosis, pneumonia and heart disease) on exposure. In addition, the health risk assessment of this pollutant as implied by the IAQ index (> 500) shows that exposure to this emission is hazardous.

Statistical analysis of emission rates data
NG emission rates data
The correlation coefficients between the pollutants released via the NG-fired kiln stack are presented in Table 2. Positive and moderate relationships are noticed to exist between CO and NO_x (0.5186), and CO and CO_2 (0.5804). Other pollutants have correlation coefficients indicating positive and weak relationships except for CO_2 and SO_x, and CO and SO_x, which have negative and weak correlations. The ANOVA test performed on the emission rates data of the pollutants from NG indicates that the data were statistically independent because $F_{observed}$ (5016.80) > $F_{critical}$ (2.8663). Also, at the 95% confidence interval, these data were found to be significant with a p-value of < 0.00001 for the ANOVA.

FO emission rates data
Correlation coefficients between the pollutants discharged from the use of FO as fuel for clinker production are provided in Table 3. It is observed that the correlation coefficients between all the pollutants showed positive-weak relationships between them. The ANOVA test conducted on the FO emission rates data indicates that the data were statistically not the same as $F_{critical}$ (2.8663) < $F_{observed}$ (3893.22). At the 95% confidence interval, the emission rates data were noticed to be significant as the p-value is < 0.00001.

3.2.3 Coal emission rates data
Table 4 gives the correlation coefficients between the pollutants emitted from using coal as fuel in the cement kiln. It is observed that only the relationship between CO_2 and NO_x is positive and relatively strong with a coefficient of 0.7476. Other correlations between the pollutants are noticed to be weak (positive and negative). ANOVA test conducted on the coal emission rates data revealed that the data were statistically not similar with $F_{critical}$ (2.8663) < $F_{observed}$ (26123.8). With a p-value < 0.00001 (at 95% confidence interval), it can be deduced that these data were significant to this study.

Table 3: Correlation of pollutants from FO.

Pollutants	NO_x	SO_x	CO_2	CO
NO_x	1			
SO_x	0.3420	1		
CO_2	0.3841	0.4809	1	
CO	0.2969	0.2030	0.2267	1

Table 4: Correlation of pollutants from coal.

Pollutants	NO_x	SO_x	CO_2	CO
NO_x	1			
SO_x	0.0098	1		
CO_2	0.7476	0.0303	1	
CO	0.2557	−0.1717	−0.0707	1

Uncertainty estimation

To estimate the uncertainties related to the emissions of CO_2, CO, SO_x, and NO_x, it was necessary to obtain suitable statistical (probability) distributions for the emission data, which were thereafter utilized to model the emissions of CO_2, CO, SO_x, and NO_x using Analytica®. EasyFit® was employed to fit the distribution for each individual pollutant using the emission data of that pollutant. The SO_x and NO_x emission data demonstrated uniform distribution while beta distribution was displayed by CO_2 and CO emission data. The values aligned to these distributions for all the pollutants were used as input variables in the input model to run the simulation by employing Analytica®. Figure 6 shows the influence diagram used in modelling the uncertainty linked to the emissions released from the cement plants. It is pertinent to know that the minimum, maximum and mean values (simulated) of the emissions obtained after running the simulation were at the 95% confidence interval. Also, the simulations were run using 500, 1000, 2000, 3000 and 4000 iterations, with the 3000th iteration giving the best result as there was no obvious change in standard deviation when compared to the 4000th iteration.

The upper and lower confidence levels of the simulated means, relative uncertainties, and means (simulated and estimated) for the release of CO_2, CO, SO_x, NO_x and total emission from the studied cement industry are provided in Table 5. In this study, the estimated means were obtained using the probability distributions subject to the probability range. In Table 5, the estimated means for all the pollutants are very close in numerical values to the simulated means. The data collection points (three kiln stacks), nature of data and their statistical distributions may be responsible for the slight differences in the mean

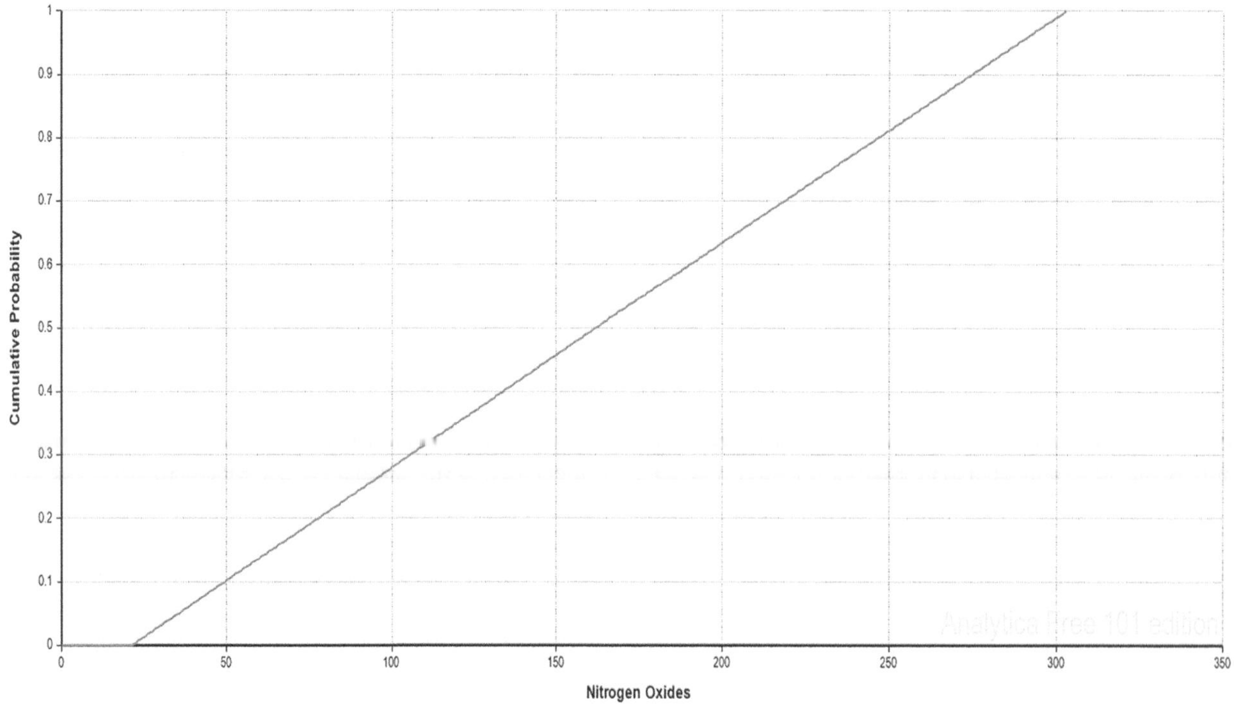

Figure 6: Cumulative distribution of nitrogen oxides emitted.

Table 5: Uncertainties associated with emissions (number of trails = 3000).

Pollutants	Min. (2.5th CL)	Simulated mean	Max. (97.5th CL)	Relative uncertainty	Estimated mean
NO_x	21.21	162	302.9	−86.91%–86.98%	162.04
SO_x	−300.2	608	1516	−149.38%–149.34%	608.01
CO_2	50.99 K	74.94 K	99.43 K	−31.96%–32.56%	75.72 K
CO	12.16	42.66	100.80	−71.50%–136.29%	39.13
Total emission	5.86 K	75.75 K	101.30 K	−32.86–33.73%	77.39 K

Note: CL = Confidence level: Negative random error = (2.5th percentile-mean)/mean; positive random error = (97.5th percentile-mean)/mean.

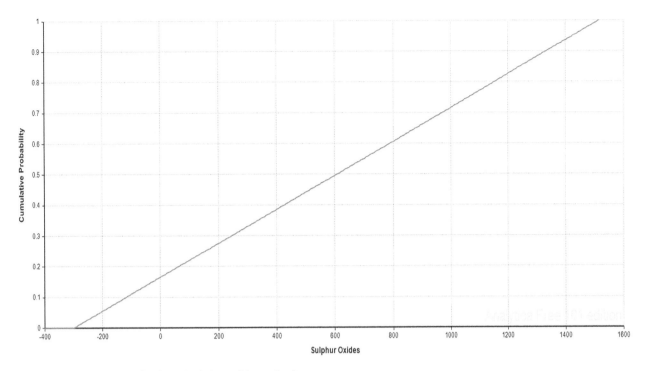

Figure 7: Cumulative distribution of sulphur oxides emitted.

values. The magnitudes of the values of the pollutants as given in Table 5 are reflections of the quantities of the pollutants discharged from the cement plants as earlier mentioned. The propagation of the uncertainties inherent in the emission data (inputs) collected for this study was used in estimating the uncertainties associated with the pollutants. As can be observed both in Table 5 and Figures 7–10, the ranges of uncertainties related to CO_2, CO, SO_x, NO_x and total emissions from the investigated

cement industry are provided. For the total emission, the estimated range of relative uncertainty was –32.86% (lower limit) to 33.73% (upper limit), corresponding to monthly simulated and estimated means of 7.58×10^4 kg and 7.74×10^4 kg, and the lower and upper limits of 5.60×10^3 kg and 1.01×10^5 kg, respectively. In addition, the relative uncertainties associated with all the pollutants (CO_2: –31.96% to 32.56%; CO: –71.50% to 136.29%; SO_x: –149.38% to 149.34%; NO_x: –86.91% to 86.98%)

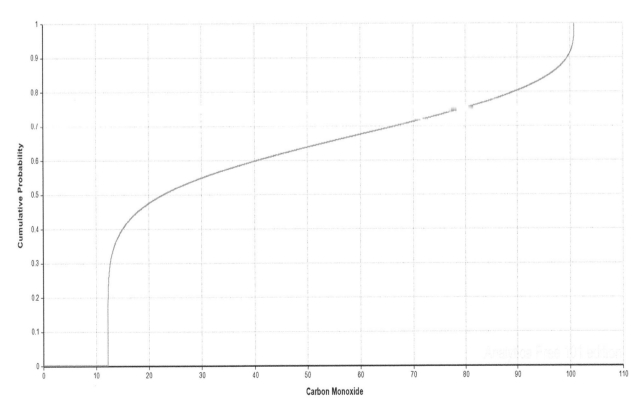

Figure 8: Cumulative distribution of carbon monoxide emitted.

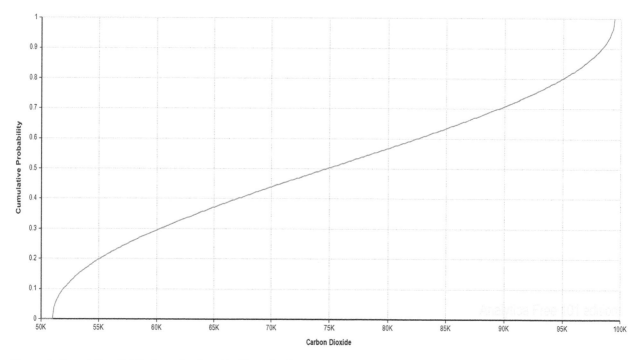

Figure 9: Cumulative distribution of carbon dioxide emitted.

were evaluated. The relative uncertainty range of 66.59% obtained for the emission of CO_2 from the cement industry is moderately higher that of 15–20% (China), 20.13–20.85% (previous study for China) and 3–5% (United States of America) reported in the literature for CO_2 uncertainty estimates of cement industries (Shen et al. 2014). The discrepancies in the uncertainty values have been primarily linked to the variations in the emission factors for the cement kilns (process-related), materials' compositions, fuels and modes of emission data collection. It is noticed that the ranges of the relative uncertainties were in the increasing order of $CO_2 < NO_x < CO < SO_x$.

Uncertainty ranges of –10% to 1000% and ± 75% have been specified for N_2O and CO_2 emissions from flared gas in developing countries, but such recommendations are not available for cement industries (Giwa et al. 2017b). The model outputs showing the cumulative distributions of CO_2, CO, SO_x, NO_x and total emissions, and indicating the minimum and maximum confidence limits are presented in Figures. 7–10.

CO_2 emissions and cement production in Nigeria

Figure 11 illustrates the quantities of cement produced and the amounts of CO_2 emission associated with them. Data

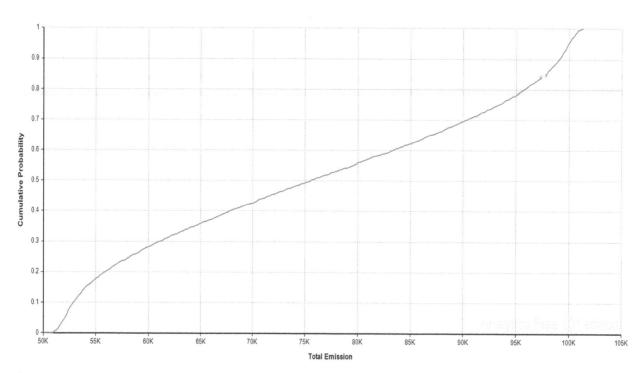

Figure 10: Cumulative distribution of total emissions released.

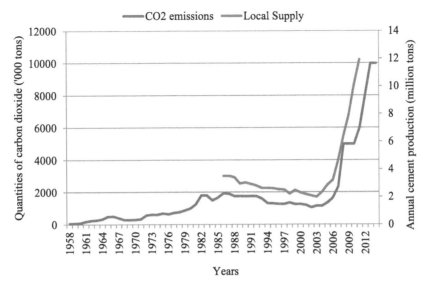

Figure 11: Carbon dioxide emission against cement production in Nigeria (Mojekwu et al., 2013; US NOAA).

on cement production and CO_2 released from cement industries in Nigeria are very scarce in the public domain, especially on the websites of national agencies saddled with this responsibility as is often the case for third world countries. As seen in Figure 11, the data on local supply (local production) of cement are only available from 1986 to 2011 (sourced from local literature) while the quantities of CO_2 released due to cement production activities in the country exist from 1959 to 2014 (United States National Oceanic Atmospheric Administration (US NOAA)). It is worth stating that the CO_2 emission data reported by NOCC for cement industries in Nigeria were collected using satellite, which is obviously a different technique compared with the method employed in this study to gather the emission rate data for the studied cement industry. There seems to be a nearly equal gap between the two lines representing the amounts of cement produced locally and quantities of CO_2 released from 1986 to 2011, except for the year 2008 (see Figure 11). This demonstrates a good relationship

between the two parameters (though from different sources) as evidenced by a high coefficient of correlation (0.965). Furthermore, the data from the two sources were insignificantly the same ($F_{critical}$ (1.9375) > $F_{obsrved}$ (0.3248) and p-value = 0.9968).

In addition, CO_2 emissions from cement production in Nigeria were compared to those of South Africa, North Africa, sub-Sahara Africa and Africa as presented in Figure 12. The CO_2 datasets used in this study were available from 1960 to 2011 and the same values were observed with those of US NOAA for Nigeria. We chose the US NOAA CO_2 data for Nigeria because it has more data (1959 to 2014) which shows the trend of CO_2 emissions for cement production in Nigeria (US NOAA 2011; DataMarket 2018). Generally, the emissions of CO_2 from cement production appear to increase considerably over 50 years for Africa and North Africa, moderately for sub-Sahara Africa, fairly for South Africa and slightly for Nigeria. A surge in CO_2 emissions for North Africa was observed to have taken place in 1976, having been

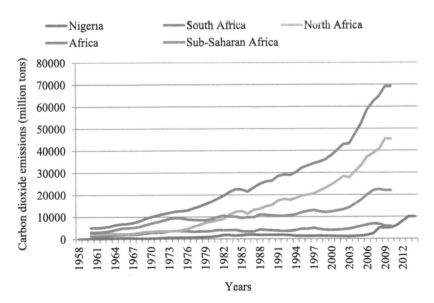

Figure 12: Trends of carbon dioxide emission from cement production (Source: US NOAA 2011; DataMarket 2018).

slightly above that of South Africa from 1960 to 1975. The amounts of CO_2 released from the cement industries in North Africa overtook that of sub-Sahara Africa from 1982 and thereafter increased progressively (see Figure 12). Comparison between Nigeria and South Africa in terms of cement production-based CO_2 emissions shows that Nigeria is fast overtaking South Africa. The trend as depicted in Figure 12 shows that the quantities of CO_2 released from Nigerian cement industries were more than those from South Africa, gradually increasing since 2007 with a possible turning point after 2011. This could be due to the establishment of additional cement industries in Nigeria and the modification of existing plants, which has substantially increased domestic cement production.

Much attention needs to be paid to the issue of emissions discharged into the environment by cement industries in Nigeria, and the attendant effects. Conducting an extensive emission inventory is paramount, as this would assist players, stakeholders and policymakers to thoroughly understand the trend and extent of emissions from this source, and the possible threat to both humans and environment, to facilitate decision making. To curb emissions from the cement industries, emissions must be monitored and reported, and mitigation strategies deployed declared, all in compliance with national emission guidelines.

Conclusion

The production of cement is mainly accompanied by the release of pollutants into the environment which is detrimental to both the microenvironment and the local environment. The emission rates of four pollutants were obtained from the utilization of three fuels in cement kilns for 7320 h of operation. This study revealed that the concentrations of the pollutants emitted into the environment were considerably higher than the recommended values. This showed that the air around the cement industry and its environment was highly polluted and could be a threat to both ecosystems and human beings. This is corroborated by the result of the health risk assessment of CO, NO_x, and SO_x, which revealed a hazardous status on exposure to the degraded air. Coal was found to emit the highest quantities of pollutants with CO_2 being the most discharged pollutants. The emission rates data for all the pollutants were found to be significant at 95% confidence interval with $F_{critical} < F_{observed}$ and p-values <0.0000001. Correlation coefficients between the pollutants for the fuels showed that only CO_2 and NO_x gave a positive and relatively strong correlation (0.7476) with others revealing weak (positive and negative) relationships between them. This study showed that the use of NG and wastes as fuel substitutes in cement kilns seems the best option to reduce the amounts of pollutants emitted, which are a continuous threat to the public health and microenvironment around the cement industry.

Acknowledgements

The contributions of Ogunbanwo Oladimeji and Adebayo Kolawole of the Department of Mechanical Engineering at Olabisi Onabanjo University, Ogun State, Nigeria, towards the success of the work are highly appreciated.

Disclosure statement

No potential conflict of interest was reported by the authors.

ORCID

Solomon O. Giwa ⓘ http://orcid.org/0000-0001-6331-2288

Collins N. Nwaokocha ⓘ http://orcid.org/0000-0002-3566-8567

References

Abdul-Wahab, S. A., G. A. Al-Rawas, S. Ali, and H. Al-Dhamri. 2016. "Assessment of Greenhouse CO_2 Emissions Associated with the Cement Manufacturing Process." *Environmental Forensics* 17 (4): 338–354. doi.10.1080/15275922.2016.1177752.

Ali, M. B., R. Saidur R, and M. S. Hossain. 2011. "A Review on Emission Analysis in Cement Industries." *Renewable and Sustainable Energy Reviews* 15: 2252–2261.

Amos, B. B., I. Musa, M. Abashiya, and I. B. Abaje. 2015. "Impacts of Cement Dust Emissions on Soils Within 10 km Radius in Ashaka Area, Gombe State, Nigeria." *Environment and Pollution* 4 (1): 29–36.

Atabi, F., M. S. Ahadi, and K. Bahramian. 2011. "*Scenario Analysis of the Potential for CO2 Emission Reduction in the Iranian Cement Industry.*" World renewable energy congress, Linkoping, Sweden, 2011, 8–13 May.

Bada, B. S., K. A. Olatunde, and A. Oluwajana. 2013. "Air Quality Assessment in the Vicinity of Cement Company." *International Research Journal of Natural Sciences* 1 (2): 34–42.

Barcelo, L., J. Kline, G. Walenta, and E. M. Gartner. 2014. "Cement and Carbon Emissions." *Materials and Structure* 47 (6): 1055–1065.

Benhelal, E., G. Zahedi, E. Shamsaei, and A. Bahadori. 2013. "Global Strategies and Potentials to Curb CO_2 Emissions in Cement Industry." *Journal of Cleaner Production* 51: 142–161.

Chatziaras, N., C. S. Psomopoulos, and N. J. Themelis. 2016. "Use of Waste Derived Fuels in Cement Industry: a Review." *Management of Environmental Quality: An International Journal* 27 (2): 178–193.

Chen, C., G. Habert, Y. Bouzidi, and A. Jullien. 2010. "Environmental Impact of Cement Production: Detail of the Different Processes and Cement Plant Variability Evaluation." *Journal of Cleaner Production* 18: 478–485.

DataMarket (now Qlik DataMarket). 2018. "CO2 emissions from Cement production (thousand metric tons)." Accessed at https://datamarket.com/data/set/1463/co2-emissions-from-cement-production-thousand-metric-tons#!ds=1463!g79=m.h.5.d.1l&display=table.

García-Segura, T., V. Yepes, and J. Alcalá. 2014. "Life Cycle Greenhouse gas Emissions of Blended Cement Concrete Including Carbonation and Durability." *International Journal Life Cycle Assessment* 19: 3–12. doi:10.1007/s11367-013-0614-0.

Giwa, S. O., A. T. Layeni, C. N. Nwaokocha, and M. A. Sulaiman. 2017b. "Greenhouse gas Inventory: A Case of gas Flaring Operations in Nigeria." *African Journal of Science, Innovation, Technology and Development* 9 (3): 241–250. doi:10.1080/20421338.2017.1312778.

Giwa, S. O., C. N. Nwaokocha, S. I. Kuye, and K. O. Adama. Forthcoming. "Gas flaring Attendant Impacts of Criteria and Particulate Pollutants: A Case of Niger Delta Region of Nigeria." *Journal of King Saud University – Engineering Sciences.* doi:10.1016/j.jksues.2017.04.003.

Giwa, S. O., C. N. Nwaokocha, and A. T. Layeni. 2016. "Assessment of Millennium Development Goal 7 in the Niger Delta Region of Nigeria via Emissions Inventory of flared gas." *Nigerian Journal Technology* 35 (2): 349–359. doi:10.4314/Njt.V35i1.1.

Hendriks, C. A., E. Worrell, D. DeJager, K. Blok, and P. Riemer. 2002. "Emission Reduction of Greenhouse Gases From the Cement Industry." In *Proceedings of the Fourth International Conference on Greenhouse Gas Control Technologies*, 939–944. Interlaken, Switzerland, 1998. https://doi.org/10.1016/b978-008043018-8/50150-8.

Ideriah, T. J. K., and H. O. Stanley. 2008. "Air Quality Around Some Cement Industries in Port Harcourt, Nigeria." *Scientia Africana* 7 (2): 27–34.

Ilalokhoin, P. O., A. J. Otaru, J. O. Odigure, A. S. Abdulkareem, and J. O. Okafor. 2013. "Environmental Impact Assessment of a Proposed Cement Plant in Southwestern Nigeria." *IOSR Journal of Environmental Science, Toxicology and Food Technology* 3 (5): 83–99.

Industry report. 2011. Nigerian Cement Industry: a review of opportunities and recurrent price hike. http://www.resourcedat.com/wp-content/uploads/2012/04/NIGERIAN-CEMENT-INDUSTRY_APRIL_2011.pdf.

IPCC (Intergovernmental Panel on Climate). 2006. IPCC Guidelines for National Greenhouse Gas Inventories, Volume 1: General Guidance and Reporting. IPCC, Geneva, Switzerland.

Lei, Y., Q. Zhang, C. Nielsen, and K. He. 2011. "An Inventory of Primary air Pollutants and CO_2 Emissions From Cement Production in China, 1990-2020." *Atmospheric Environment* 45: 147–154.

Mojekwu, J. N., A. Idowu, and O. Sode. 2013. "Analysis of the Contribution of Imported and Locally Manufactured Cement to the Growth of Gross Domestic Product (GDP) of Nigeria (1986–2011)." *African Journal of Business Management* 7 (5): 360–371.

Nigeria's Second National Communication under the United Nations Framework Convention on Climate Change. 2014. http://unfccc.int/resource/docs/natc/nganc2.pdf, accessed in June 2015.

Oyinloye, M. A. 2015. "Environmental Pollution and Health Risks of Residents Living Near Ewekoro Cement Factory, Ewekoro, Nigeria." *International Journal of Environmental, Chemical, Ecological, Geological and Geophysical Engineering* 9 (2): 108–114.

Pudasainee, D., J. Kim, J, S. Lee, S. Cho, G. Song, and Y. Seo. 2009. "Hazardous air Pollutants Emission Characteristics From Cement Kilns co-Burning Wastes." *Environ. Eng. Res* 14 (4): 212–219.

Rim-Rukeh, A. 2015. "An Assessment of Indoor Air Quality in Selected Households in Squatter Settlements, Warri, Nigeria." *Advances in Life Sciences* 5 (1): 1–11. doi:10.5923/j.als.20150501.01.

Shen, L, Cheng S, Gunson A. J., Wan H. 2005. "Urbanisation, Sustainability and the Utilization of Energy and Mineral Resources in China." *Cities* 22: 287–302. doi:10.1016/j.cities.2005.05.007.

Shen, L., T. Gao, J. Zhao, L. Wang, L. Wang, L. Liu, F. Chen, and J. Xue. 2014. "Factory-level Measurements on CO_2 Emission Factors of Cement Production in China." *Renewable and Sustainable Energy Reviews* 34: 337–349.

Srujan, S. M. 2014. "Measures to Contain Pollution Caused Due to Cement Productions: A Review." *International Journal of Emerging Technology and Advanced Engineering* 4 (11): 135–140.

US NOAA (US National Oceanic Atmospheric Administration). 2011. "National CO_2 Emissions from Fossil-Fuel Burning, Cement Manufacture, and Gas Flaring: 1751–2008." Oak Ridge: Carbon Dioxide Information Analysis Center. http://cdiac.esd.ornl.gov/ftp/trends/emissions/ngr.dat

Uson, A. A., A. M. Lopez-Sabiron, G. Ferreira, and E. L. Sastresa. 2013. "Uses of Alternative Fuels and raw Materials in the Cement Industry as Sustainable Waste Management Options." *Renewable and Sustainable Energy Reviews* 23: 242–260.

WHO (World Health Organization). 2000. *Air Quality Guidelines for Europe*. (WHO Regional Publications, European Series, No. 91, 2nd edition) Geneva: WHO.

WHO (World Health Organization). 2010. *World Health Organization (WHO) Guidelines for Indoor Air Quality: Selected Pollutants*. (The Regional Office for Europe). Geneva: WHO.

Modelling the effects of exposure to risk on junior faculty productivity incentives under the academic tenure system

Ibrahim Niankara ⓘ

This paper relies on modelling tools from probability theory and the economics of risk to analyze the ex-ante incentive properties of tenure from the perspective of junior faculty members. The theoretical results show that under publication value uncertainty and risk aversion, a junior faculty member publishes at a point where the expected value of publication exceeds the marginal cost of publication. In addition, under decreasing absolute risk aversion (DARA), increasing a junior faculty member's base salary reduces the implicit cost of private risk bearing thereby stimulating scientific productivity. However, increasing levels of uncertainty in the value of publication reduces faculty research incentives. These results have important implications for academic departments as they seek to enact effective policies to achieve and maintain their accreditation and reputation goals through maximum faculty productivity.

Introduction

Today more journals in any particular academic research field are published than anyone can reasonably keep up with. As a consequence, many articles – both print and electronic – remain without a single citation for several years. This situation, as suggested by Mohamed (2004), is simply the result of the growing 'publish-or-perish' emphasis of many academic institutions. In fact, with the 'up-or-out' rules that come with tenure-track appointment in institutions where scientific research output is the main and objective measure for tenure and promotion decisions, the emphasis on publishing introduces some degree of risk for faculty members and triggers a behavioral response in terms of scientific productivity.

Indeed, in investigating the at-work allocation of time among teaching, research, grant writing and service by science and engineering faculty at top US research universities Link, Swann, and Bozeman (2008) found that tenure and promotion have heterogeneous effects on faculty time allocation decisions depending on the faculty member's chosen career path. Ehrenberg, Pieper, and Willis (1998) found that low tenure probabilities among newly hired faculty induce higher effort and thus lead to higher faculty productivity, and ultimate tenure promotion. This result is further supported by Kou and Zhou (2009) and suggests that universities, by offering up-or-out contracts with probationary periods and predetermined academic criterion, can ensure that professors produce knowledge through research activities.

Formally, up-or-out contracts are arrangements between a department and a faculty with the following features: (i) the department commits to retain the faculty for a pre-specified period; (ii) the faculty is considered for promotion only at the end of the probationary period, subject to satisfactory completion of some departmental criteria. If promoted, permanent retention is granted, otherwise the faculty member is permanently fired (Ehrenberg and Zhang 2005). Therefore, the prospect of risk is prevalent during a junior faculty's probation period.

Freeman (1977) offers a risk-sharing explanation for the existence of the tenure system. In his opinion, the combination of the tenure system and the minimum wage policy is a risk-sharing mechanism encouraging risk-averse faculty to do risky but socially beneficial research projects. For McKenzie (1996), academic tenure is intended to guarantee the right to academic freedom, to allow original ideas to arise by giving scholars the intellectual autonomy to investigate the issues about which they are passionate, and to report unbiased findings. While Carmichael (1988) argues that tenure provides older faculty members with the needed security to select new members of potentially greater ability, McKenzie (1996) on the other hand believes that what incumbent faculty are really seeking is protection from their colleagues in a work environment operating under the rules of academic democracy. It does so by increasing the costs that predatory faculty members must incur to be successful in having more productive colleagues dismissed.

Brown (1997), on the other hand, adopts an efficiency-based explanation for the tenure system. He assumes tenure to be part of a broader system of organizational governance, where the non-profit status of the university requires faculty members to evaluate and monitor both university administrators and trustees. Therefore, tenure contracts provide incentives for faculty to behave as residual claimants, willing to assume the roles normally associated with ownership without fear or reprisal from trustees and administrators. This view is further supported by Curnalia and Mermer (2018) who believe it is in institutions' best interests to formally protect and informally promote faculty voice by formally protecting tenure and shared governance, as this would reduce turnover, increase productivity, and help institutions respond better to broader changes linked to disruptive technological advancements. On the other hand, looking at the tenure effect on teaching quality at the University of California, San Diego, Cheng (2015) found that tenure does not have a significant impact on student ratings of faculty teaching quality, at least in the immediate years after advancement for assistant and associate professors. A similar result is reported by Figlio et al. (2015) using student-level data from Northwestern University.

Song (2008) and Chen and Lee (2009), using a signalling perspective, show that tenure-track appointments play the role of a screening device by screening out low productivity faculty before the tenure contract is signed. From the perspective of junior faculty members seeking tenure, such screening is risky because the knowledge specialization required for scientific progress puts researchers at risk of being misunderstood, and not correctly evaluated by other colleagues, especially in the short run (McPherson and Winston 1983).

Whereas a normative stand is taken in other studies to justify the existence of the tenure system, the current paper follows a rather positive approach. Considering the fact that many academic institutions adopt the tenure system as an internal policy across the world, I describe junior faculty behaviour in this system to show how sources of uncertainty can affect scientific research output decisions.

This analysis is a follow-up to Chen and Lee (2009), who model the ex-ante incentives produced by academic tenure under asymmetric information using a self-selection principal-agent model with unobservable type and action and examined the incentives in different institutions. As a principal-agent model, this analysis focuses on the interest of the principal (academic institution or department), rather than that of the agent (faculty). Furthermore, none of the above referenced literature on the topic analyze the tenure system from the perspective of the faculty member seeking tenure. Therefore, the current study, which is extracted from my doctoral thesis (Niankara 2011), combines tools from the physics literature with modelling tools from probability theory and the economics of risk to provide a theoretical treatment of the ex-ante incentive properties of tenure from the perspective of junior faculty members.

The modelling strategy implemented in this paper is the first of its kind on the topic and is motivated by the idea that scientific production can be viewed as any other form of production. To this end, I start by first motivating risk under the academic tenure system in the next section, and then I describe a way to quantify faculty scientific research output in the section that follows. Thereafter is a section in which I analyze faculty publication decisions within the framework of a representative publication outlet, with uncertainty in the monetary value of publication. The penultimate section extends the sources of uncertainty to include a quantity uncertainty, in addition to the monetary value uncertainty, while the final section concludes the analysis.

Motivating risk under the academic tenure system

Risk is found in any situation where an event is not known with certainty (Chavas 2004, 5). By this definition, prospects for risk are widespread since the occurrence of any future event is almost always uncertain. In most academic institutions, where publication has become an imperative endeavour for the survival and prosperity of faculty members, the 'up-or-out' rules that usually come with tenure-track appointment in such institutions introduces risk because of the long probationary period and the inability of faculty to fully control the publication process. Usually, a well-established senior scholar is offered a tenured position directly, whereas a junior faculty with an uncertain academic prospect has to undergo a probationary period, at the end of which, either tenure will be obtained if specific academic criteria have been met or the incumbent will be fired.

The prospect of being fired introduces an overall income uncertainty for the junior faculty member. Therefore, consistent with the expected utility hypothesis by VonNeumann and Morgenstern (1944), I assume that faculty make research output decisions on the basis of the expected utility of this uncertain income. This overall income uncertainty can be motivated by two sources of uncertainty: a price uncertainty and a quantity uncertainty. To see how, let I represent the junior faculty member's overall income, then it can be represented as the sum of a (nonrandom) base salary w, and the (random) incremental income py from publishing y research output each with monetary value p. Under such formulation, the income $I = w + py$ is also a random variable because of the randomness in the second term to the right of the equality. Randomness in this second term has two possible sources. It can be introduced through uncertainty in the monetary value of publication p, or through uncertainty in the publication of scientific research output y. For this reason, the analysis presented in this paper focuses on these two sources of tenure-track risk:

(i) Uncertainty in the monetary value of publication: because of the specialized and highly sophisticated nature of academic work, its valuation is somewhat uncertain. According to McKenzie (1996), the benefits of scientific research projects undertaken by faculty are uncertain and sometimes may not be known for a long time. Moreover, the value may change with the passage of time, stressing the temporal dimension of risk. Therefore, the use of probability theory as a formal structure describing and representing risky events allows us to model the monetary value of publication in this analysis as a random variable. For example, in a study of faculty compensation, Broder and Ziemer (1982) found that an additional American Journal of Agricultural Economics (AJAE) article published every other year increases faculty salary by $735/year. In addition, some universities adopt the policy of rewarding publications with a lump sum payment for each paper published in ranked/indexed journals. Therefore, publication has a monetary value attached to it, which as a random variable, implies that it can lead to different outcomes.

(ii) Publication uncertainty: because scientific research output production is the main and objective measure considered for tenure promotion, any randomness in the publication process introduces uncertainty in the final tenure outcome. In addition, the use of the h-index described in the next section as a measure of scientific productivity shows how publication uncertainty can affect faculty probability of getting tenure after the probationary period.

The h-index is computed using the number of publications and the number of citations received by those publications; therefore, the prospect of rejection by publishers and the lack of control over research output citations introduce uncertainty in the final tenure outcome. The randomness in the publication process is modelled using a stochastic publication function in later sections.

Faculty scientific research output quantification

The h-index as adapted in this section was introduced in the physics literature by Hirsch (2005) as a useful index for characterizing the scientific output of researchers. Because junior faculty have a limited amount of time to prove their competence as researchers, this index allows for objective judgment of the impact and relevance of the faculty research work by the end of the probation period. Letting the probationary period set by the department be n years, then junior faculty publication records include among other things the number N_p of papers published over the n years, the number of citations N_c^j for each paper j, the journals where the papers were published and the journal impact parameters. A junior faculty is said to have index h if at the end of the probationary period, h of the faculty's N_p papers has at least h citations each, and the other $(N_p - h)$ papers have no more than h citations each. Hirsch (2005) argues that two faculty with similar number of total papers or total citation counts and very different h, the one with the higher h value is likely to be a more accomplished researcher. Therefore, the h index measures the broad impact of the faculty's work and avoids disadvantages of the other single-number criteria commonly used to evaluate scientists' research output such as the total number of papers N_p, the total number of citations $N_{c,tot}$, citations per paper, number of significant papers, etc.

To understand the h index, consider the case where faculty publishes n papers every year, and each publication receives c citations per year every subsequent year. The total number of citations when the tenure decision is being made, that is at the $(n + 1)th$ year, is given by

$$N_{c,tot} = \sum_{j=1}^{n} pcj = ((pcn(n + 1))/2).$$ If all papers up to year y contribute to the index, then we have

$$(n - y)c = h$$
$$py = h.$$

where the left hand sides of the above two equations represent respectively the number of citations from the most recent of the papers contributing to h and the total number of papers contributing to h. Combining those two equations yields $h = ((c)/1 + c/p) \; n \approx m \cdot n$, where $m = ((c)/1 + c/p)$. The total number of citations is approximately $N_{c,tot} \approx ((1 + (c/p))^2/2c/p) \cdot h^2 = a \cdot h^2$, where $a = ((1 + (c/p))^2/2c/p)$ therefore, the total number of citations received during the probationary period is proportional to the squared index value.

The linear relationship between the index h and the probationary period n should hold generally for junior faculty producing papers of similar quality at a steady rate over the course of the probationary period; however, the slope m will vary from faculty to faculty, and so provides a useful measure for faculty comparison. This linear relationship beaks down however when the researcher slows down in paper production or stops publishing altogether. In such a case, a stretched exponential model may be more realistic as suggested by Hirsch (2005). Since the current analysis focuses on untenured risk-averse faculty's productivity, it is reasonable to assume that faculty will not stop publishing during the probationary period, such that the linear relationship between h and n holds as a realistic model. As suggested by Hirsch (2005), a department may set a minimal value of h that a junior scholar must achieve by the end of the probationary period to secure tenure.

Without loss of generality, this analysis assumes a junior faculty chooses the highest achievable quality level publication outlet to submit all his/her research output for publication.[1] In a representative publication outlet setting, the faculty faces two types of publication uncertainty, which implicitly introduce risks in faculty likely tenure. These are: (i) uncertainty linked to the value of a publication, and (ii) uncertainty in the quantity of publication. The effects of these two sources of uncertainty on faculty behaviour are now discussed in detail.

Monetary value only uncertainty under representative outlet

First, we assume a competitive market for academic employment, with faculty research output targeted at a representative journal in a faculty's chosen field, such that publications have the same but uncertain monetary value. In the publication process, the faculty chooses the input vector $x = (x_1, \ldots, x_n)'$, which includes time, effort, and other resources used in the production of scientific research output. Faculty research output in terms of publications is denoted by y, and the publication technology represented by the function $y = f(x)$. The publication function $f(x)$ measures the largest feasible research output the faculty member can obtain by committing the input vector $x = (x_1, \ldots, x_n)'$. At this point, I assume no uncertainty in the publication process itself (such a case will be discussed in later sections). At the time, research project decisions are made, the faculty tries to anticipate the uncertain monetary value he/she will receive from publishing his/her research output. As such, faculty treats the monetary value of publication p as a random variable, with a given subjective probability distribution.

Let $v = (v_1, \ldots, v_n)'$ denote the respective prices paid for the inputs $x = (x_1, \ldots, x_n)'$. Then the faculty cost of producing research output can be represented by $v' = \sum_{i=1}^{n} v_i x_i$, and the uncertain income generated is py. It follows that faculty net monetary benefit from publication can be represented by $\tau = py - v'x$. In addition, letting w denotes the base salary or initial wealth, then faculty terminal wealth is $w + py - v'x$. Given that the monetary value of publication is uncertain, this terminal wealth is also uncertain. Now assuming faculty behave in a way consistent with the expected utility hypothesis, then the objective function of the faculty member is

$$E[U(w + py - v'x)] = E[U(w + \tau)] \qquad (1)$$

where the E is the expectation operator based on the subjective probability distribution of the random variable p. It's assumed that faculty have risk-averse preferences represented by the utility function $U(\cdot)$ which satisfies $U' \equiv \partial U / \partial w > 0$ and $U'' \equiv \partial 2 \ U / \partial w^2 < 0$. To see how the utility function $U(\cdot)$ summarizes all risk information relevant to faculty decisions, consider the following assumptions about faculty preferences among risky prospects b_1 and b_2 where,

$b_1 \sim {}^*b_2$ implies indifference between b_1 and b_2
$b_1 \geq {}^* b_2$ implies that b_2 is not preferred to b_1
$b_1 > {}^* b_2$ implies that b_1 is preferred to b_2

ASSUMPTION A1 *(Ordering and transitivity)*

- For any random variables b_1 and b_2, exactly one of the following must hold: $b_1 > {}^*b_2$, $b_2 > {}^*b_1$ or $b_1 \sim {}^*b_2$.

  If $b_1 \geq {}^*b_2$ and $b_2 \geq {}^*b_3$ then $b_1 \geq {}^*b_3$. (Transitivity)

ASSUMPTION A2 *(Independence)*

For any random variables b_1, b_2, b_3, and any $\alpha(0 < \alpha < 1)$, then $b_1 \leq {}^* b_2$ if and only if $[\alpha b_1 + (1-\alpha)b_3] \leq {}^* [\alpha b_2 + (1-\alpha)b_3]$.

(That is, the preferences between b_1 and b_2 are independent of $b3$)

ASSUMPTION A3 *(Continuity)*

For any random variables b_1, b_2, b_3, where $b_1 < {}^* b_3 < {}^* b_2$, there exist numbers α and β, $(0 < \alpha < 1)$, $(0 < \beta < 1)$, such that $b_3 < {}^* [\alpha b_2 + (1 - \alpha)b_1]$ and $b_3 > {}^* [\beta b_2 + (1 - \beta)b_1]$.

(That is, sufficiently small change in probabilities will not reverse a strict preference)

ASSUMPTION A4 For any risky prospects b_1, b_2 satisfying $P\ r[b_1 \leq r : b_1 \leq {}^* r] = P\ r[b_2 \geq r : b_2 \geq {}^*r] = 1$ for some sure reward r, then $b_2 \geq {}^* b_1$.

ASSUMPTION A5 *(Ordering and transitivity)*

- For any number r, there exist two sequences of numbers $\alpha_1 \geq {}^* \alpha_2 \geq {}^* \dots$ and $\beta_1 \leq {}^* \beta_2 \leq {}^* \dots$ Satisfying $\alpha m \leq {}^* r$ and $r \leq {}^* \beta n$ for some m and n.
- For any risky prospects b_1 and b_2, if there exists an integer m_o such that

  $[b_1$ conditional on $b_1 \geq \alpha_m : b_1 \geq {}^* \alpha_m] \geq {}^* b_2$ for every $m \geq m_o$, then $b_1 \geq {}^* b_2$, and if there exists an integer n_o such that $[b_1$ conditional on $b_1 \leq \beta_n : b_1 \leq {}^* \beta_n] \leq {}^* b_2$ for every $n \geq n_o$, then $b_1 \leq {}^* b_2$.

Under assumptions A1–A5, for any risky prospects b_1 and b_2, by the expected utility theorem, there exists a utility function $U(b)$ representing faculty risk preferences such that $b_1 \geq {}^*b_2$ if and only if $E[U(b_1)] \geq {}^* E[U(b_2)]$, with $U(b)$ defined up to a positive linear transformation.

An extended proof of the expected utility theorem is provided in DeGroot (1970, 113–114).

Therefore, under the assumption A1–A5, the expected utility hypothesis provides an accurate characterization of faculty behaviour under risk in general and, as such, can be used to model faculty response to risk during the probation period in tenure-track appointment.

Result 1. The definition of the utility function, up to a positive linear transformation, implies that, if $U(b)$ is a utility function for a particular faculty, then so is the linear transformation $Z(b) = + \beta U(b)$ for any α and $\beta > 0$ scalars.

Proof: Starting from the equivalence between $b_1 \geq {}^* b_2$ and $E[U(b_1)] \geq {}^* E[U(b_2)]$, stated in the expected utility theorem, given $\beta > 0$, $E[U(b_1)] \geq {}^* E[U(b_2)]$ is equivalent to $\alpha + \beta E[U(b_1)] \geq \alpha + \beta E[U(b_2)]$, which is also equivalent to $E[Z(b_1)] \geq E[Z(b_2)]$. Therefore, $b_1 \geq {}^* b_2$ if and only if $E[Z(b_1)] \geq E[Z(b_2)]$, implying that $Z(\cdot)$ and $U(\cdot)$ provide equivalent representations of a faculty member's risk preferences.

This characteristic further implies that without affecting a faculty member's preference ranking, the utility function $U(b)$ can be shifted by changing its intercept and/or by multiplying its slope by a positive constant. This special feature will be useful for our analysis.

Now, letting $\mu = E(p)$ be the expected monetary value of publication, then p can be represented as $p = + \sigma e$, where e is a random variable with zero mean, and introducing randomness in the monetary value of publication p. The random component e can exhibit any distribution for which both the mean and the variance exist. The standard deviation of the monetary value of publication, σ, can be interpreted as a mean-preserving spread parameter for the distribution of p. Therefore, in this analysis, the probability distribution of p will be characterized by the mean μ and the mean preserving spread parameter σ following the analysis of firm production under uncertainty by Sandmo (1971).

Under the expected utility model, a faculty member's publication decision can be represented by:

$$\text{Max}_{xy} \{E[U(w + py - v'\ x) : y = f(x)]\}. \qquad (2)$$

which states that publication decisions are made in a way consistent with expected utility maximization. Given this objective function in Equation (2) we can now proceed to analyze faculty cost minimizing behaviour, publication function, and behavioural responses to changes in key decision variables.

Faculty costs minimizing behaviour

In the absence of uncertainty in the quantity of publication, expected utility maximization implies cost minimization by faculty. Faculty will minimize the cost of the inputs used in the publication process, which includes the effort cost, the opportunity cost of time allocated to research, and the costs of other used resources. To see how, note that the maximization problem in Equation (2) can be written as:

$$\text{Max}_y\{\text{Max}_x\{E[U(w + py - v'x):y = f(x)]\}\}$$
$$= \text{Max}_y\{E[U(w + py + \text{Max}_x\{-v'x:y = f(x)\}]\}$$
$$= \text{Max}_y\{E[U\ w + py - \text{Min}_x\{v'x:y = f(x)\}]\} \quad (3)$$
$$= \text{Max}_y\{E[U(w + py - C(v, y)]\}.$$

where $C(v, y) = \text{Min}_x\{v'x:\ y = f(x)\}$ is the publication cost function similar to the cost function in standard production theory under certainty. Therefore, for given input prices v, $C(v, y)$ measures the smallest possible cost of producing research output y, where output here is measured in terms of publications. This shows that in the absence of publication risk, risk-averse junior faculty has the incentive to behave in a cost-minimizing fashion.

Equation (3) offers a convenient way of analyzing a junior faculty member's behaviour, since it involves choosing only one variable: y, faculty scientific research output. Now assuming that the scholar decides on positive research output, $y > 0$, then using the chain rule, the first-order necessary condition associated with the optimal choice of y is given by:

$$F(y, \cdot) \equiv E[U' \cdot (p - C_y)] = 0, \quad (4)$$

or recalling from rules of probability theory that $Cov(U', p) = E(U' \cdot p) - E(U' \cdot E(p)$, then $E(U' p) = E(U') \cdot \mu + Cov(U', p)$, and Equation (4) can be rewritten as:

$$\mu - C_y + Cov(U', p)/EU' = 0 \quad (5)$$

where $C_y \equiv \partial C/\partial y$ denotes the marginal cost of publication, $U \equiv \partial U/\partial y$, and
$Cov(U', p) = E(U'\sigma e)$. The associated second order condition for a maximum is

$$D \equiv \partial F/\partial y \equiv E[U(-C_{yy})] + E[U(p - C_y)^2] < 0 \quad (6)$$

Define R to be the Arrow-Pratt measure of risk premium, which shows the shadow cost of private risk bearing by faculty. Then R is the monetary value satisfying the indifference relationship $\{w + py - C(v, y)\} \sim^* \{w + E(p) \cdot y - C(v, y) - R\}$ and under the expected utility model R is the solution of the equation:

$$E[U[w + py - C(v, y)]] = U[w + E(p) \cdot y - C(v, y) - R] \quad (7)$$

So, given that $U[w + py - C(v, y)]$ is a strictly increasing function, its inverse always exists. Denoting the inverse by U^{-1}, then it follows that $U^{-1}\{E[U[w + py - C(v, y)]]\} = w + E(p) \cdot y - C(v, y) - R$, thus the risk premium can be written as

$$R(w, y, \cdot) = w + \mu \cdot y - C(v, y) - U^{-1}\{E[U[w + py - C(v, y)]]\}. \quad (8)$$

So, maximizing the expected utility as shown in Equation

(3) is equivalent to maximizing 'the Certainty equivalent' $w + \mu \cdot y - C(v, y) - R(w, y, \cdot)$. It follows then that faculty publication decision can alternatively be written as:

$$\text{Max}_y[w + \mu \cdot y - C(v, y) - R(w, y, \cdot)], \quad (9)$$

with the associated first-order condition given by

$$\mu - C_y(v, y) - R_y(w, y, \cdot) = 0 \quad (10)$$

where $R_y(w, y, \cdot) \equiv \partial R/\partial y$ is the marginal risk premium. Comparing this result with the first order condition in Equation (5), it follows that $Ry(w, y, \cdot) = -Cov(U', p)/E(U')$, providing an intuitive interpretation for the covariance term: $[-Cov(U', p)/E(U')]$ as the marginal risk premium, measuring the marginal effect of scientific research output production, on the implicit cost of private risk bearing by faculty. Resolving the first order condition in Equation (10) above gives the publication function $y^*(w, \mu, \sigma)$ as presented below.

The publication function
The publication function is the function $y^*(w, \mu, \sigma)$ satisfying the first-order condition in Equation (5), or in Equation (10):

$$\mu = C_y(v, y) + R_y(w, y, \cdot) \quad (11)$$

The condition in Equation (11) implies that, at the optimum research output publication y^*, the expected monetary value of publication μ is equal to the marginal cost of publication C_y, plus the marginal risk premium R_y. This means that expressing the sum $(C_y + R_y)$ as a function of research output y gives the publication function, and generates the schedule of scientific research output production by the risk-averse faculty member, for each level of expected monetary value of publication μ.

Result 2. Under uncertainty in the monetary value of publication, and risk aversion, a junior scholar publishes at a point where the expected monetary value of publication μ exceeds the marginal cost of publication C_y.

Proof: if $\partial U'/\partial p < 0(>0)$, then U' and p move in the opposite(same) direction(s), implying a negative (positive) covariance, therefore the covariance term $Cov(U', p)$ is always of the sign of $\partial U'/\partial p$. But sign $[\partial U'/\partial p] = $ sign $(U'' \cdot y)$. Thus, risk aversion (where $U'' < 0$) implies that $Cov(U', p) < 0$. And it follows that the marginal risk premium $R_y = -Cov(U', p)/E(U') > 0$ under risk aversion, which in turn implies that $\mu > C_y$ at the optimum.

The publication function is illustrated in Figure 1, and shows that under risk aversion, risk can have significant effects on a faculty member's resource allocation. The analysis shows that, while risk does not involve any explicit cost to faculty, its implicit cost (as measured by the marginal risk premium R_y) needs to be added to the marginal cost of publication C_y in the evaluation of optimal publication decisions. Since risk affects faculty scientific productivity under risk aversion, I now investigate this effect in more details by conducting a comparative static analysis of the research output decision y in Equation (4).

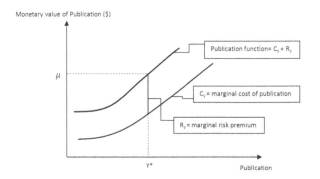

Figure 1: Publication function.

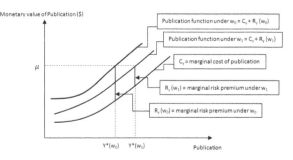

Figure 2: Effect of changing base salary w under DARA, with $w1 > w0$.

Comparative statics analysis

Letting $\alpha = (w, \mu, \sigma)$ be the vector of parameters of the publication function $y^*(\alpha)$, then using the chain rule and total differentiating the first-order condition $F(y, \alpha) = 0$ at the optimum $y = y^*(\alpha)$ yields

$$\partial F/\partial\alpha + (\partial F/\partial y)(\partial y*/\partial\alpha) = 0,$$

or, with $D = \partial F/\partial y < 0$, (from Equation (6))

$$\partial y*/\partial\alpha = -D^{-1}\partial F/\partial\alpha$$
$$= -D^{-1}\partial\{E[U' \cdot (p - C_y)]\}/\partial\alpha$$
$$= \text{sign}(\partial\{E[U' \cdot (p - Cy)]\}/\partial\alpha).$$

This result is used to analyze the properties of the publication function $y^*(\alpha)$, looking at changes in the elements of the parameter vector $\alpha = (w, \mu, \sigma)$.

The effect of a change in base salary, w

The effect of changing faculty base Salary (initial wealth) w is given by

$$\partial y*/\partial w = -D^{-1}\{\partial\{E[U^0 \cdot (p - C_y)]\}/\partial w\}$$
$$= -D^{-1}\{E[U^{00} \cdot (p - C_y)]\}. \quad (12)$$

Assuming faculty preferences exhibit decreasing absolute risk aversion (DARA), then the term $E[U'' \cdot (p - C_y)] > 0$. To see that, consider the Arrow-Pratt absolute risk aversion coefficient $r = -U''/U'$., and any risky return b then the result $R \approx -0.5(U''/U') \cdot Var$ (b) by (Chavas 2004, 36–37) provides a link between the risk premium R and the Arrow-Pratt coefficient of absolute risk aversion $r = -U''/U'$. Because $var(b) > 0$ for all b non degenerate, the sign of the risk premium R always is the same as that of r, and so for risk-averse faculty $(R > 0)$, the corresponding $r = -U''/U' > 0$. Therefore, letting τ_o denote the net benefit from publication τ, when evaluated at $p = Cy$, then under DARA,

$$r(\tau) \overset{<}{>} r(\tau_o) \quad \text{if} \quad p \overset{>}{<} C_y,$$

it follows that

$$-U''/U' \overset{<}{>} r(\tau_o) \quad \text{if} \quad p \overset{<}{>} C_y,$$

or

$$U'' \overset{<}{>} -r(\tau_o) \cdot U' \quad \text{for} \quad (p - C_y) \overset{>}{<} 0,$$

or

$$U'' \cdot (p - C_y) > -r(\tau_o) \cdot U' \cdot (p - C_y),$$

and taking the expectation on both sides of the inequality yields

$$E[U'' \cdot (p - C_y)] > -r(\tau_o) \cdot E[U' \cdot (p - C_y)] = 0,$$

with the last equality on the RHS coming from the first order conditions in Equation (4). Therefore

$$E[U'' \cdot (p - C_y)] > 0 \text{ as required and } \partial y^*/\partial w > 0.$$

Result 3. Under uncertainty in the monetary value of publication, if DARA characterizes faculty preferences, increasing a junior faculty member's base salary w tends to stimulate scientific research output production by decreasing the implicit cost of private risk bearing. This occurs because under DARA, private wealth accumulation and insurance motives are substitutes.

This result is illustrated in Figure 2 and shows that, under DARA, increasing the faculty member's base salary w from w_0 to w_1 reduces the marginal risk premium $R_y(w)$. This is because under risk aversion and DARA, private wealth accumulation reduces the risk premium R, which is accompanied by a reduction in the marginal risk premium $R_y(w)$, which further generates a shift to the right of the publication function $(C_y + R_y)$.

The effect of a change in expected monetary value of publication μ

The effect of changing expected monetary value of publication μ is given by

$$\partial y*/\partial\mu = -D^{-1}\{\partial\{E[U^0 \cdot (p - C_y)]\}/\partial\mu\}$$
$$= -D^{-1}\{E[U^0 + yE[U^{00} \cdot (p - C_y)]\}. \quad (13)$$

Now defining $\partial y^c/\partial\mu \equiv -D^{-1}[E(U')]$ as the compensated expected monetary value effect, and given that $D < 0$ (from the second-order condition in Equation (6)), it follows that $\partial yc/\partial\mu > 0$. This in turn implies that the 'compensated' publication function is always upward sloping with respect to changes in the expected monetary value of publication μ. Also, recalling from Equation (12) that $\partial y^*/\partial w = -D^{-1}\{E[U'' \cdot (p - C_y)]\}$, then we have the following slutsky equation:

$$\partial y^*/\partial \mu = \partial y^c/\partial \mu + (\partial y^*/\partial w) \cdot y^*, \qquad (14)$$

which suggests that the slope of the uncompensated monetary value of publication $\partial y^*/\partial \mu$ is equal to that of the compensated monetary value $\partial y c/\partial \mu$, plus an income effect $(\partial y^*/\partial w) \cdot y^*$. Under DARA, it was shown in result 3 that $(\partial y^*/\partial w) > 0$; therefore, the income effect $(\partial y^*/\partial w) \cdot y^*$ is also positive. Given that the compensated effect is $\partial y c/\partial \mu > 0$, from the slutsky equation it follows that $\partial y^*/\partial \mu > 0$. This implies that the publication function exhibits a positive slope with respect to the uncompensated monetary value of publication.

Result 4. Under uncertainty in the monetary value of publication, if DARA characterizes faculty preferences, then an increase in expected monetary value of scientific research output stimulates a junior faculty member's publication effort.

This result is illustrated in Figure 3 and shows that under DARA increasing the expected monetary value of research output μ from μ_0 to μ_1 increases publications $y^*(\mu_1) > y^*(\mu_0)$, since the publication function is upward sloping under DARA. Also, Figure 3 suggests that the marginal cost C_y and the marginal risk premium R_y are higher at μ_1 compared to μ_0.

The effect of a change in publication value risk σ

The effect of changing the mean-preserving parameter value σ, which also represents the risk in the monetary value of publication, is given as follows, with a standardized value of sigma ($\sigma = 1$)

$$\partial y^*/\partial \sigma = -D^{-1}\{\partial\{E[U' \cdot (p-C_y)]\}/\partial \sigma\},$$
$$= -D^{-1}\{E(U' \cdot e) + yE[U'' \cdot (p - \mu)(p - C_y)]\},$$
$$= -D^{-1}\{E(U' \cdot e) + yE[U'' \cdot (p-C_y + C_y - \mu)(p - C_y)]\},$$
$$= -D^{-1}\{E(U' \cdot e) + yE[U'' \cdot (p - C)^2] + y(C_y - \mu)E[U'' \cdot (p - C_y)]\}.$$
$$(15)$$

But $E(U' \cdot e) = Cov(U', p)$, since $p = + \sigma e$ with $E(e) = 0$. In addition,

$Cov(U', p) = sign(U'', y) < 0$ under risk aversion, and $(U'' < 0)$ implies $E[U'' \cdot (p - C_y)^2] < 0$.

Furthermore, it was shown in results (2) and (3) that under DARA, $C_y - \mu < 0$ and $^E[U'' \cdot (p - Cy)] > 0$. It follows then from Equation (15), that $\partial y^*/\partial \sigma < 0$ that is, increasing publication value risk (σ), decreases optimal research output.

Result 5. Under risk aversion, if DARA characterizes junior faculty preferences, for a given mean return to publication, an increase in publication value risk provides a general disincentive to publish.

This result is illustrated in Figure 4, where increasing σ from σ_0 to σ_1, shifts the publication function to the left. This increasing risk exposure increases the scholar's private risk bearing as measured by the risk premium R, and so increases the marginal risk premium R_y. The publication function $(C_y + R_y)$ increases such that for a given value of publication, research output falls.

Monetary value and quantity uncertainty under representative outlet

So far, our analysis has focused on faculty behaviour under uncertainty in the monetary value of publication alone. The sources of uncertainty are now extended by introducing the second source of tenure-track risk in the form of uncertainty in the quantity of publication, where risk is also introduced through faculty research output, rather than just the value of the output itself. This is motivated by the fact that while faculty have the power to make choices over the inputs committed in the publication process, the outcome of the process as to whether or not the journal accepts the research output for publication is beyond the faculty member's control. To this end, we consider a general setting of such uncertainty, followed next by special cases for practical considerations.

The general setting of quantity of publication uncertainty

Under general publication uncertainty, a junior faculty member's research output is a random variable at the time input choices are being made. The publication technology can be represented by a stochastic publication function. A generic specification of this function is $y(x, e)$, where y is the research output, x is a vector of inputs, and e is a random variable reflecting publication uncertainty.

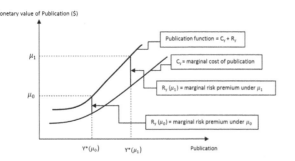

Figure 3: The effect of changing mean value of publication μ under DARA, with $\mu 1 > \mu 0$.

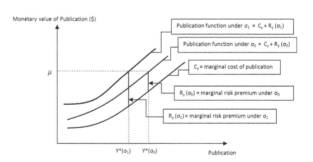

Figure 4: The effect of changing publication value risk σ under DARA, with $\sigma 1 > \sigma 0$.

The stochastic publication function $y(x, e)$ provides the maximum possible research output that can be obtained when the input vector x is committed and the random variable e is realized. The faculty member is assumed to have information about publication uncertainty, information represented by a subjective probability distribution of the random variable e.

Building on price uncertainty from the previous section, let p be the monetary value of the faculty member's research output, v, represent the prices of the inputs used in the publication process, and w the base salary received by faculty. Then, the income generated by the junior scholar's publications is $py(x, e)$, the cost of publication is $v' = \sum_{i=1}^{n} v_i x_i$, the net return from publication is $\tau = py(x, e) - v'x$, and faculty terminal income (wealth) is $(w + \tau)$. Further, allowing for uncertainty in the monetary value of publication, then the faculty member does not know either e or p at the time research input decisions are being made. In such a setting, the faculty member faces both price and publication risks and so treats both e and p as random variables with a given subjective joint probability distribution. Under the expected utility model, the junior faculty's objective function is to choose inputs (time, effort, and other resources used in the publication process) so as to maximize expected utility of terminal income

$$\text{Max}_x \{E[U(w + py(x, e) - v'x)]\} \qquad (16)$$

where the expectation E is based on the joint subjective distribution of the random variables (p, e). Using the chain rule, the first order necessary conditions for the optimal choice of inputs is:

$$E[U' \cdot (p\partial y(x, e)/\partial x - v)] = 0,$$

or

$$E[(p\partial y(x, e)/\partial x] = v - Cov[U', p\partial y(x, e)/\partial x]/E(U'),$$

or

$$E(p)E[\partial y(x, e)/\partial x] + Cov[p, \partial y(x, e)/\partial x]$$
$$= v - Cov[U', p\partial y(x, e)/\partial x]/E(U'). \qquad (17)$$

As shown in Equation (9), maximizing the expected utility is equivalent to maximizing the corresponding certainty equivalent. The certainty equivalent of faculty terminal income (wealth) is
$w + E[py(x, e)] - v'x - R(x, \cdot)$, where $R(x, \cdot)$ represents the Arrow-Pratt risk premium as before. Therefore, the maximization problem in Equation (16), can be rewritten as:

$$\underset{\leq x \leq}{\text{Max}} \{w + E[py(x, e)] - v'x - R(x, \cdot)\}$$

with first-order conditions

$$\partial E[py(x, e)]/\partial x - v - R_x(x, \cdot) = 0,$$

or

$$\partial E[py(x, e)]/\partial x = v + R_x(x, \cdot). \qquad (18)$$

$R_x(x, \cdot) \equiv \partial R(x, \cdot)/\partial x$ represents the marginal risk premium. Comparing the first-order conditions in Equation (17) to Equation (18) indicates that the marginal risk premium takes the form:
$R_x(x, \cdot) = -Cov[U', p\partial y(x, e)/\partial x]/E(U')$. This equality allows us to characterize the covariance term $-Cov[U', p\partial y(x, e)/\partial x]/E(U')$ as the marginal risk premium measuring the effects of the committed input vector x, on the implicit cost of private risk bearing by the junior faculty member. It also shows that at optimal input commitment by the junior faculty, the expected marginal value of research output, $\partial E[py(x, e)]/\partial x$, is equal to the per unit cost of committed input v plus the marginal risk premium, $R_x(x, \cdot)$.

In the general form of the stochastic publication function $y(x, e)$, the marginal risk premium can be either positive, negative, or zero. Whether a particular input increases or decreases the implicit cost of risk faced by the faculty member is largely an empirical matter. However, for risk-averse faculty, the risk premium $R(x, \cdot) > 0$ and so when the marginal risk premium $R_{x_i}(x, \cdot) > 0$, a faculty member will have an incentive to reduce the use of the i-th input because this input increases the implicit cost of risk bearing. For example, when the i-th input is 'effort' exerted in the research process, then a positive marginal risk premium $R_{x_i}(x, \cdot) > 0$ provides the junior scholar with the incentive to reduce such 'effort'. Conversely, when $R_{x_i}(x, \cdot) < 0$, the i-th input reduces the implicit cost of risk, and so gives the junior faculty member the incentive to increase the use of this input in the publication process. For example, if the i-th input is 'time', then $R_{x_i}(x, \cdot) < 0$ gives the junior scholar the incentive to put more time into the research activity. For empirical tractability, we now consider possible specifications for the general form of the publication function $y(x, e)$.

The Just-Pope specification of the stochastic publication function

Just and Pope (1978, 1979) propose flexible specifications of stochastic production functions in general. Because of the parallelism between the publication function developed in this paper and the standard stochastic production function, a similar specification is adopted for the stochastic publication function, which can be represented as:

$$y(x, e) = f(x) + e[h(x)]^{1/2}, \qquad (19)$$

where $E(e) = 0$ and $Var(e) > 0$. This specification of the publication function implies that $E(y) = f(x)$ and $\partial E(y)/\partial x = \partial f(x)/\partial x$, and that $Var(y) = Var(e)h(x)$ with $\partial Var(y)/\partial x = Var(e) \cdot \partial h(x)/\partial x$. Since the $var(e) > 0$, the sign of $\partial Var(y)/\partial x$ depends on that of $\partial h(x)/\partial x$.

The publication function as specified can be interpreted as a regression model exhibiting heteroscedasticity, where the interest is on identifying the effects of faculty input decisions on the variance $Var(y)$ of the research output produced. A higher research output variance effect implies a riskier input choice. In this Just-Pope like specification of the publication function, depending on the functional form of $h(x)$, the marginal risk premium $\partial h(x)/\partial x$ can be negative, positive, or zero. Therefore, inputs used in the publication process will be identified as risk reducing, risk increasing, or risk neutral. In situations where the inputs affect publication risk, $\partial h(x)/\partial x = 0$, faculty can manage risk exposure through judicious choice of inputs. Under risk aversion, where $R(x,\cdot) > 0$, faculty have an incentive to use inputs which reduce risk exposure and its implicit cost and for which $\partial h(x)/\partial x < 0$. In such a setting, risk has a direct impact on faculty input allocation and thus publication decisions.

The moment-based specification of the stochastic publication function

A moment-based approach, which includes the Just-Pope specification as a special case, is adapted from Antle (1983). The Just-Pope specification, also referred to as mean-variance specification, is based on only the first two moments of the distribution of the stochastic publication function and, as such, is a special case of the more general moment-based approach. The moment-based approach allows for the empirical exploration of the role of higher-order moments and can capture more interesting features of the stochastic publication function. Considering the stochastic publication function in its generic form $y(x, e)$, then the moment generating function (MGF) of the random research output y, if it exists, is given by $M_y(t) = E(exp\ (ty))$. The r-th derivative of the MGF evaluated at $t = 0$ gives the r-th moment about the origin $M_Y^r(0) = E(Y^r)\ \forall\ r = 1,\ 2,$. from which the first central moment, the mean of the publication function given the committed input vector x, is obtained by setting $r = 1$. Let the mean be denoted as:

$$\mu(x) = E[y(x, e)],$$

then the r-th moment about the mean of the stochastic publication function is given by:

$$M_r(x) = E\{[y(x, e) - \mu(x)]^r\}\forall r = 2, 3, \ldots \quad (20)$$

From Equation (20) we have $M_2(x) = Var(x)$, $M_3(x)$, and $M_4(x)$ respectively as the conditional variance, skewness, and kurtosis of the faculty member's stochastic publication function.

The skewness $M_3(x)$ for example, provides a measure of symmetry of the distribution and so can be adjusted to accommodate various forms of risk aversion. To make this approach empirically tractable, the following two specifications can be used:

$$y = \mu(x) + u\ \forall\ r = 1 \quad (21)$$

from which $[y - \mu(x)] = u$. Raising both sides of the equality to the r-th power gives the second specification as:

$$[y - \mu(x)]^r \equiv u^r = M_r(x) + v_r, \ \forall\ r = 2, 3, \ldots \quad (22)$$

where $E(u) = E(v_r) = 0$ and $Var(u) = M_2(x)$ while

$$Var(v_r) \equiv E[u^r - M_r]^2 = E(u^{2r}) + M_r^2 - 2E(u^r)M_r$$
$$= M_{2r} - M_r^2.$$

As specified, Equations (21) and (22) are standard regression models that can be implemented empirically in a non-parametric (Hayfield and Racine 2008) or semi-parametric (Ruppert, Wand, and Carroll 2009) estimation framework. One can also specify a parametric form for $\mu(x)$ and $M_r(x)$ and use generalized method of moments estimation(GMM) (Baum, Schaffer, and Stillman 2003), weighted least squares (WLS), or ordinary least squares with heteroscedasticity consistent covariance matrix estimator (OLS with HCCME) (Zeileis, 2004) for estimation. All these estimators are consistent while accounting for the non-constant variance (heteroscedasticity). Because of space constraints, and for the sake of brevity, the discussion is limited to only the theoretical results; the empirical treatment will be fully detailed in a subsequent research article.

Conclusion

The growing emphasis on publication as a requirement for tenure affects junior faculty members' scientific productivity incentives by raising the uncertainty level they face. This paper has attempted to provide a theoretical justification of the effect of such risks on faculty scientific research output decisions, using a perspective not yet considered in the literature. It was shown that a risk-averse faculty member in the presence of publication value uncertainty publishes at a point where the expected value of publication exceeds its marginal cost. In addition, a faculty member's scientific productivity was shown to be stimulated by an increase in base salary when decreasing absolute risk aversion (DARA) described faculty preferences. Conversely, increased uncertainty in the value of publication provided a lesser incentive for scientific research output production.

In terms of faculty ex-ante behaviour under the tenure system, the results from this analysis are consistent with those obtained under the principal agent-model with asymmetric information (Song 2008), (Chen and Lee 2009). Tenure track contracts with up-or-out rules introduce risks that significantly influence junior faculty productivity incentives. In this analysis, although information about faculty ability is not known to the department/university, a department through its ranking and publication quality standards, indirectly sends information about tenure requirements. Such information is taken into account by risk-averse junior faculty looking to secure a permanent matching with their hiring departments.

In the principal agent model, tenure track as a labour market contract acts as a screening device, providing departments with information on faculty ability. In this case, it reduces the information rent for the department, and operates only after the tenure-track contract is signed. However, In the current analysis tenure contracts with up-or-out rules regulate faculty self-selection mechanism into departments even before the contract is signed, since risk affects faculty members' decisions on which departments to seek tenure contracts from based on perceived subjective probability of survival in the department. Once a choice is made and tenure-track contract signed, then tenure rules continue to operate affecting faculty scientific productivity incentives as they identify the highest ranked publication outlet in their field that meets departmental quality requirement for promotion.

Finally, It must be emphasized that this study provides a purely positive analysis and is intended to describe the effect of risk on junior faculty publication incentives as faculty themselves might perceive it. Nevertheless, a detailed examination of this sort seems worthwhile to illuminate such a familiar and commonplace phenomena from a rather different angle, using already established tools from probability theory and the economics of risk. In order to keep the exposition concise, i only cover here the case of homogeneous research output with a unique but uncertain value. The heterogeneous case, along with the empirical implementation are left out for future extension.

Disclosure statement

No potential conflict of interest was reported by the author.

Note

1. Following the philosophy of 'representative agent', i defines a 'representative publication outlet' to refer to the typical journal of a certain type/ranking (for example, the typical A ranked, or B ranked journal, by official ranking institutions such as the Australian Business Deans Council (ABDC) Journal quality list), with a given monetary value. This is grounded on the assumption that on average, a faculty member publishes all research output in journals with the highest achievable quality ranking subject to his or her research ability, such that we can reasonably aggregate all publications into a single average quality ranking with a unique monetary value for each member.

ORCID

Ibrahim Niankara http://orcid.org/0000-0001-7049-0059

References

Antle, J. M. 1983. "Testing the Stochastic Structure of Production: A Flexible Moment-based Approach." *Journal of Business and Economics Statistics* 1: 192–201.

Baum, C. F., M. E. Schaffer, and S. Stillman. 2003. "Instrumental Variables and GMM: Estimation and Testing." *Stata Journal* 3 (1): 1–31.

Broder, J. M., and R. F. Ziemer. 1982. "Determinants of Agricultural Economics Faculty Salaries." *American Journal of Agricultural Economics* 64: 301–303.

Brown Jr W. O. 1997. "University Governance and Academic Tenure: A Property Rights Explanation." *Journal of Institutional and Theoretical Economics (JITE)/Zeitschrift für die gesamte Staatswissenschaft* 153 (3): 441–461

Carmichael, H. L. 1988. "Incentives in Academics: Why is There Tenure?" *Journal of Political Economy* 96 (3): 453–472.

Chavas, J.-P. 2004. *Risk Analysis in Theory and Practice*. San Diego, CA: Elsevier, Inc.

Chen, Z., and S.-H. Lee. 2009. "Incentives in Academic Tenure Under Asymmetric Information." *Economic Modelling* 26 (2): 300–308.

Cheng, D. A. 2015. "Effects of Professorial Tenure on Undergraduate Ratings of Teaching Performance." *Education Economics* 23 (3): 338–357.

Curnalia, R. M. L., and D. Mermer. 2018. "Renewing our Commitment to Tenure, Academic Freedom, and Shared Governance to Navigate Challenges in Higher Education." *Review of Communication* 18 (2): 129–139.

DeGroot, M. H. 1970. *Optimal Statistical Decisions*. New York: McGraw-Hill.

Ehrenberg, R. G., P. J. Pieper, and R. A. Willis. 1998. "Do Economics Departments with Lower Tenure Probabilities Pay Higher Faculty Salaries?" *Review of Economics and Statistics* 80 (4): 503–512.

Ehrenberg, R. G., and L. Zhang. 2005. "Do Tenured and Tenure-track Faculty Matter?" *Journal of Human Resources* 40 (3): 647–659.

Figlio, D. N., Schapiro, M. O., & Soter, K. B. 2015. "Are tenure track professors better teachers?." *Review of Economics and Statistics* 97(4): 715–724.

Freeman, S. 1977. "Wage Trends as Performance Displays Productive Potential: A Model and Application to Academic Early Retirement." *The Bell Journal of Economics* 8 (2): 419–443.

Hayfield, T., and J. S. Racine. 2008. "Nonparametric Econometrics: The np Package." *Journal of Statistical Software* 27 (5). http://www.jstatsoft.org/v27/i05/.

Hirsch, J., November 2005. "An Index to Quantify an Individual's Scientific Research Output." *Proceedings of the National Academy of Sciences of the United States of America* 102 (56): 16569–16572.

Just, R. E., and R. D. Pope. 1978. "Stochastic Specification of Production Functions and Economic Implications." *Journal of Econometrics* 7: 67–86.

Just, R. E., and R. D. Pope. 1979. "Production Function Estimation and Related Risk Considerations." *American Journal of Agricultural Economics* 61: 276–284.

Kou, Z., and M. Zhou. October 2009. Multi-tasking vs. Screening: A Model of Academic Tenure. *Accessed June 2009*. http://ssrn.com/abstract=1483049.

Link, A. N., C. A. Swann, and B. Bozeman. 2008. "A Time Allocation Study of University Faculty." *Economics of Education Review* 27 (4): 363–374.

McKenzie, R. B. 1996. "In Defense of Academic Tenure." *Journal of Institutional and Theoretical Economics* 152: 325–341.

McPherson, M. S., and G. C. Winston. 1983. "The Economics of Academic Tenure: a Relational Perspective." *Journal of Economic Behavior and Organization* 4 (2): 163–184.

Mohamed, G. March 2004. Publish or Perish—An Ailing Enterprise? *Physics Today* A/4/4/4/44 June 2003, http:// www.people.vcu.edu/gadelhak/Opinion.pdf.

Niankara, I. L. C. O. 2011. "Essays in Risk and Applied Bayesian Econometric Modeling" Doctoral diss., Oklahoma State University. Accessed October 22, 2018. https://shareok.org/bitstream/handle/11244/6721/Department20of20Economics20and20Legal%20Studies_13.pdf?isAllowed=y&sequence=1.

Ruppert, D., M. P. Wand, and R. J. Carroll. 2009. "Semiparametric Regression During 2003–2007." *Electronic Journal of Statistics* 3: 1193–1256.

Sandmo, A. 1971. "On the Theory of Competitive Firm Under Price Uncertainty." *American Economic Review* 61: 65–73.

Song, J. 2008. "Tenure and Asymmetric Information: An Analysis of an Incentive Institution for Faculty Development in Research Universities." *Frontiers of Education in China* 3 (2): 310–319.

VonNeumann, J., and O. Morgenstern. 1944. *Theory of Games and Economic Behavior*. Princeton: Princeton University Press.

Zeileis, A. 2004. "Econometric Computing with HC and HAC Covariance Matrix Estimators." *Journal Of Statistical Software, Articles* 11 (10): 1–17. https://www.jstatsoft.org/v011/i10.

Analyses of mathematical models for city population dynamics under heterogeneity

O.C. Collins, T.S. Simelane and K.J. Duffy

In this study, a mathematical model which takes heterogeneity into account is presented to describe city population dynamics. Initial insight is gained by qualitative analyses of a homogeneous version of this model. To understand the population dynamics of South African cities the model is applied to case studies of three major cities, Cape Town, Durban and Johannesburg. For these cases, parameter estimations are calculated from data extracted from population statistics provided for South Africa from our own and other research. The relative importance of the parameters is investigated through sensitivity analyses. Overall, it is shown that a full understanding of the population dynamics, important for city planners, should consider income and expenditure heterogeneity of these South African cities.

Introduction

Current urbanization worldwide provides opportunities and challenges for socioeconomic development and global sustainability (Brelsford et al. 2017; Montgomery 2008; UN-Habitat 2012). Urban areas generate most of the global economy (Solecki, Seto, and Marcotullio 2013). Cities hold more than half of the world's population, consume over half of the world's energy, and emit much of the world's pollution (Solecki, Seto, and Marcotullio 2013). Thus, understanding and developing mechanisms that generate improved living conditions and economic growth for all cities are important (Brelsford et al. 2017). These developmental efforts are necessary for effective planning and implementation of urban policies.

African has one of the world's fastest rates of urbanization (Kihato 2018). For example, more than half of South African people are urbanized (Kok and Collinson 2006). While the effects of urbanization on communities can be fairly neutral (Hatton and Tani 2005), they can also be negative or positive. Examples of positive effects include that urban areas can benefit from the increased labour and intellectual capital and people with new jobs can send resources home to rural areas (Kok and Collinson 2006). The negative effects of urbanization can include a number of factors such as increased diseases (Rees et al. 2010; Vearey et al. 2010), a decline in services and education, increased unemployment (Kok and Collinson 2006), increased poverty, political instability (Arieff 2010), food insecurity (Crush and Frayne1 2007), a lack of housing, and stresses placed in general on urban space developments (Balbo and Marconi 2005).

These positive and negative effects of urbanization can have complex results on city dynamics. Thus, understanding the dynamics and drivers of African urbanization are important. The processes and actual numbers involved in urban growth dynamics are important for urban planners, and regional and national policymakers, as well as for the individuals involved. In this study computational modelling is used to understand the dynamics better. In particular, we show how heterogeneous aspects in the factors involved in urbanization can have important consequences on the overall population dynamics of African cities.

It is widely accepted that there are factors, or drivers, that result in urbanization (Van Hear, Bakewell, and Long 2018). For example, income has been shown to be a primary driver for migrating to cities (Rapanos 2005; Harris and Todaro 1970; Zenou and Smith 1995). Income-triggered city migration can then have an important influence on the population dynamics of a transforming city (Odularu 2014; Onwe 2013). In African, cities are generally experiencing growth in middle income earners (Deloitte 2014; Gounden and Nkhumeleni 2013) and this results in increased expenditure. As one might expect, economic growth leads to increased urbanization and the transformation of cities (Epstein et al. 1967; Todaro 1997). This study uses mathematical models to investigate the influence of income, and the associated expenditure, on city population dynamics. This work hopes to assist city planners. In particular, in highlighting heterogeneous population effects such models could be important in planning infrastructural developments. For example, for certain cities more schools and universities might be needed as compared to other cities.

Heterogeneity has been considered extensively in studying the dynamics of infectious diseases. For instance, Robertson, Eisenberg, and Tien (2013) investigated heterogeneity of multiple transmission pathways of waterborne diseases. Collins and Govinder (2014) used similar models to study waterborne cholera disease in Haiti. Even though most cities are made up of heterogeneous populations, the concept of heterogeneity has not been fully explored. The aim of this study is to consider city population dynamics using a mathematical model that includes the necessary heterogeneity.

The remaining parts of this work are organized as follows. A mathematical model that describes a city's population dynamics is formulated in the next section. To gain insight into the overall dynamics of this model, qualitative analyses of a special case are carried out in the section thereafter. Using the model, population dynamics of three major cities in South Africa are presented in the penultimate section. We conclude the work by discussing our results in the final section.

Model formulation

In this section, we consider the features of city populations and formulate a mathematical model that can describe the dynamics. Most city adult populations can minimally be seen as comprising five sub group categories (communities): workers, job seekers, business owners, students and visitors. Visitors are other people such as immigrants or tourists who do not fit into one of the other categories. These groups may overlap slightly but this is ignored. People who fit none of these categories are taken to be dependents. Each of these communities can be represented as a function of time t. Let the city populations of a working community, business community, student community, visitor's community and job seeking community, with their dependents, at time t be denoted by $W(t)$, $B(t)$, $S(t)$, $V(t)$ and $J(t)$ respectively. Therefore, the total city population at time t denoted by $N(t)$ is $N(t) = W(t) + B(t) + S(t) + V(t) + J(t)$.

People earn different incomes and to take different income levels into account we assume that all the individuals on a particular income level form a homogeneous sub population of each particular community. For instance, let S_i be a sub population of the student community S at a particular income level, then $S = \sum_{i=1}^{n} S_i$, where n is the total number of different income levels. For the remaining communities, let (B_i, J_i, W_i, J_i) be homogeneous sub populations with the same income level of business people (B), job seekers (J), workers (W) and visitors (V) such that $B = \sum_{i=1}^{n} B_i$, $J = \sum_{i=1}^{n} J_i$, $W = \sum_{i=1}^{n} W_i$ and $V = \sum_{i=1}^{n} V_i$. The subscript i is used to emphasize the income level, i.e. income level i. Since what a person can spend is determined by the person's income, we assume that each subgroup is categorized by individuals that fall within the same income and expenditure level. Based on these assumptions, we note that there are two types of dynamics within the city: the global and internal dynamics. Part of the internal dynamics is between income levels which represents a form of heterogeneity necessary for a proper description. Considering heterogeneity in this way is more realistic and should give a better representation of population dynamics of any city.

In the model, the populations of the communities- $(S_i, B_i, J_i, W_i, V_i)$ are assumed to change over time. We assume that individuals at income level i enter the student community (S_i) at a rate Λ_i^S and upon graduation some go into business at a rate β_i, some go seeking for job at a rate σ_i, while others will leave the city at a rate λ_i^S. Some people will enter a city for business purposes while some will relocate for economic or financial reasons. Thus, we assume that individuals are recruited into the business community at a rate Λ_i^B and also leave the business community at a rate λ_i^B. Another primary reason that people go to cities is for employment. For this factor, we assume that individuals enter the working community W_i at a rate Λ_i^W or leave (due to retirement, loss of a job or death) at a rate λ_i^W. Note that some people migrate to cities for work since they are perceived as having (or have) more jobs but these people do not always find work. Thus, job seekers increases at a rate Λ_i^J but those finding a job do so at rate σ_i while those who could not find a job leave the city at a rate λ_i^J. Job seekers can also interact with business people with the aim of learning how to start up a business (entrepreneurs). We assume that interaction between job seekers and business people increases entrepreneurship among job seekers at a rate φ_i. Finally, visitors increase at rate Λ_i^V and leave the city at a rate λ_i^V. Visitors obtain jobs at a rate δ_i or start studying at a rate γ_i. Note that communities also have dependents and relatives. Thus, the recruitment terms are assumed to cover birth, death and movements in and out of the city.

Income and expenditure are expected to be major factors of migration. To understand the effects of income and expenditure on city population dynamics, it is necessary to incorporate these in our model. We assume that cities with more income attract individuals and those with less do not. Let $(I_i^S, I_i^B, I_i^J, I_i^W, I_i^V)$ be measures of attraction due to potential income by each of the communities at income level i. Also, let $(E_i^S, E_i^B, E_i^J, E_i^W, E_i^V)$ be measures of repulsion due to expenditure potential or lack of income of each community.

For the internal dynamics, workers can be promoted or get a better paying job or lose their job. Thus, workers in income level i move to income level j as their income changes at rate w_{ij}. Similarly, business people in income level i move to income level j as their income changes through making more profit or loss at rate b_{ij}. Since students, job seekers and visitors in any income level do not have the opportunity of making more income, we assume that they remain in their respective income levels until they finish their studies. Putting these formulations together we obtain the equations

$$\frac{dS_1}{dt} = \Lambda_1^S + \gamma_1 V_1 - (\beta_1 + \sigma_1 + \lambda_1^S + E_1^S - I_1^S)S_1,$$

$$\frac{dB_1}{dt} = \Lambda_1^B + \beta_1 S_1 + \varphi_1 B_1 J_1 - (\lambda_1^B + E_1^B - I_1^B)B_1$$
$$- \sum_{i=1}^{n} b_{1i}B_1 + \sum_{i=1}^{n} b_{i1}B_i,$$

$$\frac{dJ_1}{dt} = \Lambda_1^J + \sigma_1 S_1 + \delta_1 V_1 - (\alpha_1 + \lambda_1^J + E_1^J - I_1^J)J_1$$
$$- \varphi_1 B_1 J_1,$$

$$\frac{dW_1}{dt} = \Lambda_1^W + \alpha_1 J_1 - (\lambda_1^W + E_1^W - I_1^W)W_1 - \sum_{i=1}^{n} w_{1i}W_1$$
$$+ \sum_{i=1}^{n} w_{i1}W_i,$$

$$\frac{dV_1}{dt} = \Lambda_1^V - (\delta_1 + \gamma_1 + \lambda_1^V + E_1^V - I_1^V)V_1,$$

$$\frac{dS_2}{dt} = \Lambda_2^S + \gamma_2 V_2 - (\beta_2 + \sigma_2 + \lambda_2^S + E_2^S - I_2^S)S_2,$$

$$\frac{dB_2}{dt} = \Lambda_2^B + \beta_2 S_2 + \varphi_2 B_2 J_2 - (\lambda_2^B + E_2^B - I_2^B)B_2$$
$$- \sum_{i=1}^n b_{2i}B_2 + \sum_{i=1}^n b_{i2}B_i,$$

$$\frac{dJ_2}{dt} = \Lambda_2^J + \sigma_2 S_2 + \delta_2 V_2 - (\alpha_2 + \lambda_2^J + E_2^J - I_2^J)J_2$$
$$- \varphi_2 B_2 J_2,$$

$$\frac{dW_2}{dt} = \Lambda_2^W + \alpha_2 J_2 - (\lambda_2^W + E_2^W - I_2^W)W_2$$
$$- \sum_{i=1}^n w_{2i}W_2 + \sum_{i=1}^n w_{i2}W_i, \qquad (1)$$

$$\frac{dV_2}{dt} = \Lambda_2^V - (\delta_2 + \gamma_2 + \lambda_2^V + E_2^V - I_2^V)V_2,$$

$$\vdots$$

$$\frac{dS_n}{dt} = \Lambda_n^S + \gamma_n V_n - (\beta_n + \sigma_n + \lambda_n^S + E_n^S - I_n^S)S_n,$$

$$\frac{dB_n}{dt} = \Lambda_n^B + \beta_n S_n + \varphi_n B_n J_n - (\lambda_n^B + E_n^B - I_n^B)B_n$$
$$- \sum_{i=1}^n b_{ni}B_n + \sum_{i=1}^n b_{in}B_i,$$

$$\frac{dJ_n}{dt} = \Lambda_n^J + \sigma_n S_n + \delta_n V_n - (\alpha_n + \lambda_n^J + E_n^J - I_n^J)J_n$$
$$- \varphi_n B_n J_n,$$

$$\frac{dW_n}{dt} = \Lambda_n^W + \alpha_n J_n - (\lambda_n^W + E_n^W - I_n^W)W_n - \sum_{i=1}^n w_{ni}W_n$$
$$+ \sum_{i=1}^n w_{in}W_i,$$

$$\frac{dV_n}{dt} = \Lambda_n^V - (\delta_n + \gamma_n + \lambda_n^V + E_n^V - I_n^V)V_n,$$

with initial conditions $S_i(0) \geq 0$, $B_i(0) \geq 0$, $J_i(0) \geq 0$, $W_i(0) \geq 0$ and $V_i(0) \geq 0$. Note that $b_{ii} = w_{ii} = 0$.

Model analyses

In this section, we present the mathematical analyses of model (1) with the aim of improving our understanding of the city population dynamics. To gain insight into the dynamics as well as determine the potential effects of income and expenditure we start our analyses by considering a homogeneous version of model (1) (i.e. by setting $n = 1$).

Homogeneous version of model (1) without income and expenditure

If we set $n = 1$ in model (1) and $E^S - I^S = E^B - I^B = E^J - I^J = E^W - I^W = E^V - I^V = 0$, we obtain

$$\frac{dS}{dt} = \Lambda^S + \gamma V - (\beta + \sigma + \lambda^S)S,$$

$$\frac{dB}{dt} = \Lambda^B \beta S - \lambda^B B,$$

$$\frac{dJ}{dt} = \Lambda^J + \sigma S + \delta V - (\alpha + \lambda^J)J, \qquad (2)$$

$$\frac{dW}{dt} = \Lambda^W + \alpha J - \lambda^{W\Lambda^W} W,$$

$$\frac{dV}{dt} = \Lambda^V - (\delta + \gamma + \lambda^V)V,$$

with initial conditions $S(0) \geq 0$, $B(0) \geq 0$, $J(0) \geq 0$, $W(0) \geq 0$, $V(0) \geq 0$.

Analysis of model (2)

Analysis of this model is fundamental since it captures the basic city population dynamics. For simplicity, we let $\mu = \delta + \gamma + \lambda^V$, $\xi = \alpha + \lambda^J$ and $\rho = \beta + \sigma + \lambda^S$. Equation (2) has an equilibrium point given by

$$(S^0, \ B^0, J^0, \ W^0, V^0)$$
$$= \left(\frac{\Lambda^S + \gamma V^0}{\rho}, \ \frac{\Lambda^B + \beta S^0}{\lambda^B}, \ \frac{\Lambda^J + \sigma S^0 + \delta V^0}{\xi}, \right. \qquad (3)$$
$$\left. \frac{\Lambda^W + \alpha J^0}{\lambda^W}, \ \frac{\Lambda^V}{\mu} \right).$$

Model (2) is solved analytically to determine its exact solution and is given by

$$V(t) = V^0 + K^V e^{-\mu t},$$

$$S(t) = S^0 + \frac{\gamma K^V}{\rho - \mu} e^{-\mu t} + K^S e^{-\rho t},$$

$$B(t) = B^0 + \frac{\beta \gamma K^V}{(\rho - \mu)(\lambda^B - \mu)} e^{-\mu t} + \frac{\beta K^S}{\lambda^B - \rho} e^{-\rho t}$$
$$+ K^B e^{-\lambda^B t}, \qquad (4)$$

$$J(t) = J^0 + \frac{\sigma \gamma K^V}{(\xi - \mu)(\rho - \mu)} e^{-\mu t} + \frac{\sigma K^S}{\xi - \rho} e^{-\rho t}$$
$$+ \frac{\delta K^V}{\xi - \mu} e^{-\mu t} + K^J e^{-\xi t},$$

$$W(t) = W^0 + \frac{\alpha\sigma\gamma K^V}{(\xi - \mu)(\rho - \mu)(\lambda^W - \mu)}e^{-\mu t}$$

$$+ \frac{\alpha\sigma K^S}{(\xi - \rho)(\lambda^W - \rho)}e^{-\rho t} + \frac{\alpha\delta K^V}{(\xi - \mu)(\lambda^W - \mu)}e^{-\mu t}$$

$$+ \frac{\alpha K^J}{\lambda^W - \xi}e^{-\xi t} + K^W e^{-\lambda^W t},$$

where

$$K^V = V(0) - V^0,$$

$$K^S = S(0) - \left(S^0 + \frac{\gamma K^V}{\rho - \mu}\right),$$

$$K^B = B(0) - \left(B^0 + \frac{\beta\gamma K^V}{(\rho - \mu)(\lambda^B - \mu)} + \frac{\beta K^S}{\lambda^B - \rho}\right),$$

$$K^J = J(0) - \left(J^0 + \frac{\sigma\gamma K^V}{(\xi - \mu)(\rho - \mu)} + \frac{\sigma K^S}{\xi - \rho} + \frac{\delta K^V}{\xi - \mu}\right),$$

$$K^W = W(0) - \left(W^0 + \frac{\alpha\sigma\gamma K^V}{(\xi - \mu)(\rho - \mu)(\lambda^W - \mu)}\right.$$

$$\left. + \frac{\alpha\sigma K^S}{(\xi - \rho)(\lambda^W - \rho)} + \frac{\alpha\delta K^V}{(\xi - \mu)(\lambda^W - \mu)} + \frac{\alpha K^J}{\lambda^W - \xi}\right),$$

are constants of integration.

Theorem 3.1 *Model (2) is globally asymptotically stable.*

Proof: The proof of **Theorem 3.1** is straight forward. From the solution of model (2), we see that

$$(S(t), B(t), J(t), W(t), V(t))$$

$$\to (S^0, B^0, J^0, W^0, V^0) \text{ as } t \to \infty. \quad (5)$$

Equation (5) implies that model (2) is globally asymptotically stable completing the proof.

An implication of **Theorem 3.1** is that the communities and the entire city population will not go to infinity irrespective of the initial population size. This agrees with basic population growth models as well as logistic population growth models in which the carrying capacity is not exceeded (Pearl and Reed 1920; Tsoularis 2001; Verhulst 1838). Another interesting implication of the theorem is that the parameters determine the size of future populations and not the initial population size of the city. For example, migration rates within and across the communities are important determinates of the final population size in this model. This agrees with other studies where migration is the major factor that influences city population dynamics (Goldstein 1990).

Homogeneous version of model (1) with income and expenditure

Here we present a homogeneous version of model (1) in the presence of income and expenditure. By setting

$n = 1$ in model (1), we obtain

$$\frac{dS}{dt} = \Lambda^S + \gamma V - (\beta + \sigma + \lambda^S + E^S - I^S)S,$$

$$\frac{dB}{dt} = \Lambda^B + \beta S - (\lambda^B + E^B - I^B)B,$$

$$\frac{dJ}{dt} = \Lambda^J + \sigma S + \delta V - (\alpha + \lambda^J + E^J - I^J)J, \quad (6)$$

$$\frac{dW}{dt} = \Lambda^W + \alpha J - (\lambda^W + E^W - I^W)W,$$

$$\frac{dV}{dt} = \Lambda^V - (\delta + \gamma + \lambda^V + E^V - I^V)V,$$

with initial conditions $S(0) \geq 0$, $B(0) \geq 0$, $J(0) \geq 0$, $W(0) \geq 0$, $V(0) \geq 0$.

Analysis of model (6)
Note that model (6) can also be solved analytically to obtain its exact solution. In fact, if we set $\hat{\mu} = \delta + \gamma + \lambda^V + E^V - I^V$, $\widehat{\lambda^B} = \lambda^B + E^B - I^B$, $\hat{\xi} = \alpha + \lambda^J + E^J - I^J$ and $\hat{\rho} = \beta + \sigma + \lambda^S + E^S - I^S$, then model (6) reduces to model (2). The solution of model (6) can easily be obtained from the solution of model (2). Thus, we will not present the solution of model (6) to avoid repetition of a similar analysis. However, comparing the solutions of model (6) with the solutions of model (2) will help us determine the effects of income and expenditure for the city population dynamics.

Effects of income and expenditure
To improve our understanding on the effects of income and expenditure the long term dynamics of the solutions of model (6) are considered. This analysis will be carried out only for the student community $S(t)$ but similar analyses can be done for other communities. The effects of income and expenditure in the population dynamics of the student community are considered for three possible cases below.

Case (i) Income greater than expenditure
If in the student community $\beta + \sigma + \lambda^S + E^S - I^S < 0$, (i.e. income far greater than expenditure), our analysis reveals that $S(t) \to \infty$. In reality $S(t)$ may not reach infinity, but this analysis shows that if student income in the form of bursaries or scholarships are greater than expenditure then student population sizes can grow very large. This could partially explain increasing populations in some cities where incomes are high or better funding for study opportunity exist.

Case (ii) Income less than expenditure
On the order hand if $\beta + \sigma + \lambda^S + E^S - I^S > 0$, (i.e. income less than expenditure), then $S(t) \to S^0$ at infinite time. That is the population of students tends to the steady state of the model (system).

Case (iii) Income and expenditure approximately equal
The final case when income and expenditure are approximately equal reverts to model (1) which we have already discussed.

These results should also hold for the other categories of business people, workers, job seekers and visitors. To clarify further, we present a case study of the three major cities in South Africa: Cape Town, Durban and Johannesburg.

Case study: Cape Town, Durban and Johannesburg
Here we present realistic case studies for the cities of Cape Town, Durban and Johannesburg. The data for the case studies are collected from two sources: primary and secondary. The primary source of data is from our own field surveys (Tables 1 and 2). The other source of data are from published data mostly from Statistics South Africa (SSA 2014). These data can be found in Table 3 and the web pages of Statistics South Africa.

By using these primary and secondary data sets, we estimated the parameter values by fitting the homogeneous model (6) that includes measures of income and expenditure for the cities of Cape Town, Durban and Johannesburg to match the population values from the data of Statistics South Africa (Table 3). The estimated parameter values (γ, β, σ, δ, α) for these cities together with the estimated measures of income and expenditure are given in Table 4. The model was found to fit the data well which shows that the model can effectively describe the population dynamics of each of these cities. Thus, the model can be used to study and predict future population dynamics of these cities for various groups. The results of these analyses are presented in Figures 1–3. These results show that the cities of Cape Town, Durban and Johannesburg could be dominated by workers and business people in the future. This possibility agrees with city population dynamics globally as work and businesses are among the major factors that attract people to cities. Next, the

Table 1: Reasons (%) for people to migrate to cities in South African.

Reasons	Cape Town	Durban	Johannesburg
Education	23%	35%	33%
Employment	70%	60%	67%
Business	7.0%	3.0%	0.0%
Poverty/Job Seekers	0.0%	1.0%	0.0%
Visit	0.0%	1.0%	0.0%

Table 2: Income categories of city residents in South Africa.

Income levels	Categories	Cape Town	Durban	Johannesburg
Income level 1	R1000 – R3000	35%	17%	8.5%
Income level 2	R3001 – R5000	18%	17%	24%
Income level 3	R5001 – R7000	22%	32%	28.5%
Income level 4	R7001 – R10000	9.0%	23%	24%
Income level 5	>R10000	16%	11%	15%

Table 3: Population numbers across South Africa's cities using Census 2001 and 2011.

Cities	2001 Census	2011 Census	Average annual growth	Source
Cape Town	2 799 496	3 740 026	2.9%	(SSA 2014)
Durban	3 091 938	3 442 361	1.1%	(SSA 2014)
Johannesburg	3 308 568	4 434 827	3.0%	(SSA 2014)

Table 4: Estimated measures of income, expenditure and parameter values for the cities.

Parameter	Symbol	Cape Town	Durban	Johannesburg
Measure of expenditure for S	E^S	0.0175	0.0024	0.0165
Measure of income for S	I^S	0.0001	0.0001	0.0001
Measure of expenditure for B	E^B	0.0001	0.0001	0.0001
Measure of income for B	I^B	0.0695	0.0135	0.0757
Measure of expenditure for J	E^J	0.0134	0.0076	0.0026
Measure of income for J	I^J	0.0004	0.0001	0.0001
Measure of expenditure for W	E^W	0.0018	0.0001	0.0001
Measure of income for W	I^W	0.0001	0.0001	0.0075
Measure of expenditure for V	E^V	0.0160	0.0001	0.0120
Measure of income for V	I^V	0.0120	0.0442	0.0130
Migration rate from V to S	γ	0.0130	0.0693	0.0100
Migration rate from S to B	β	0.0052	0.0378	0.0212
Migration rate from S to J	σ	0.1246	0.0546	0.0324
Migration rate from V to J	δ	0.0160	0.0010	0.0120
Migration rate from J to W	α	0.7825	0.0885	0.0500

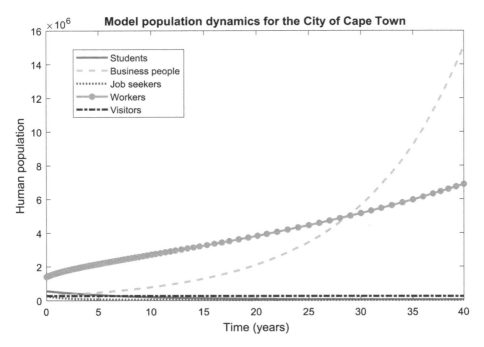

Figure 1: Graphical illustration of population dynamics for various communities (i.e. students, visitors, job seekers, business people, workers) in the city of Cape Town.

same parameter estimates are used to investigate the total population dynamics of the three cities (Figure 4). From the figure, the total population for each of these cities is on the increase with Johannesburg having the greatest population growth rate followed by Cape Town and then Durban.

To determine the impacts of income and expenditure on the population dynamics of the cities, model (6) is solved numerically. The numerical solution for $(S(t), B(t), J(t), W(t), V(t))$ in the presence or absence of a measure of income and expenditure for these three cities is presented in Figures 5–7. For Cape Town (Figure 5) including factors for income and expenditure results in a decrease in students, job seekers, workers and the visitor population, but an increase in the business population and the total population of the city. For the city of Durban (Figure 6), including factors for income and expenditure results in an increase in students, visitors, business people and the total population, but a decrease in workers and the job seeker population. For the city of Johannesburg (Figure 7), including factors for income and expenditure results in an increase in business people, visitors, workers and the total population, but a decrease in students and job seekers. Thus, considering

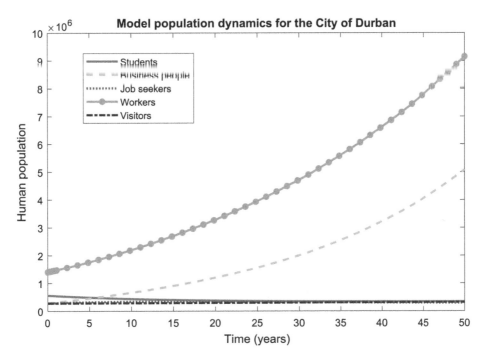

Figure 2: Graphical illustration of population dynamics for various communities (i.e. students, visitors, job seekers, business people, workers) in the city of Durban.

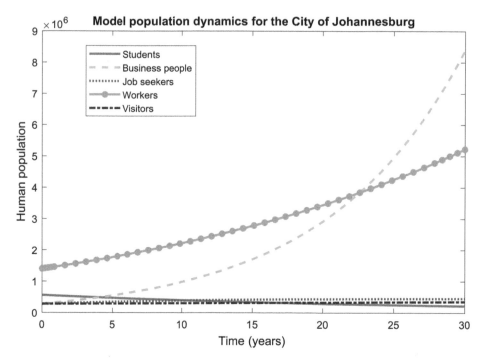

Figure 3: Graphical illustration of population dynamics for various communities (i.e. students, visitors, job seekers, business people, workers) in the city of Johannesburg.

income and expenditure can have varied impacts on these groups (students, workers, visitors, business people, and job seekers) for different cities. For instance, considering income and expenditure has negative impacts on the student populations of Cape Town and Johannesburg but a positive impact on the student population of Durban.

Sensitivity analysis

To understand the relative importance of the different parameters (factors) responsible for city population dynamics, it is useful to carry out a sensitivity analysis of model (6). We calculate sensitivity indices for model

(6) to reveal the relative importance of each parameter on the population dynamics.

Definition 4.2 (Chitnis, Hyman, and Cushing 2008). The normalized forward sensitivity index of a variable u that depends on a parameter p is defined as:

$$Y_p^u = \frac{\partial u}{\partial p} \times \frac{p}{u}. \qquad (7)$$

When $Y_p^u > 0$, we say that p increases the value of u as its value increases, while if $Y_p^u < 0$ we say that p decreases the value of u as its value increases.

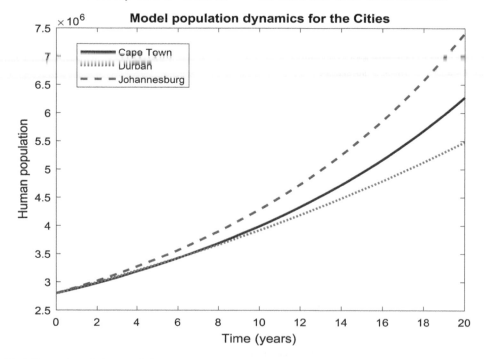

Figure 4: Graphical illustration of city population dynamics for the city of Cape Town, Durban and Johannesburg.

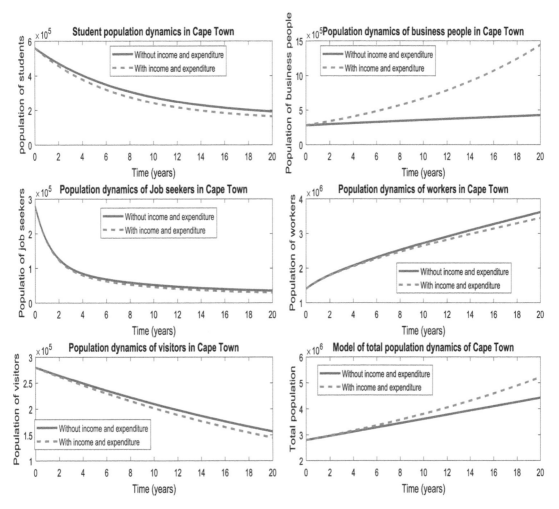

Figure 5: Graphical illustration of the effects of income and expenditure on the population dynamics for various communities (i.e. students, visitors, job seekers, business people, workers) in the city of Cape Town.

Since we have an explicit solution of model (6), we can derive an analytical expression for the sensitivity indices of these solutions with respect to each of the parameters using the above definition. To determine the magnitude of the sensitivity indices we use the estimated parameter values in Table 4. This method of calculating sensitivity indices has been successfully implemented for population model dynamics of infectious diseases (Hove-Musekwa et al. 2011; Mwasa and Tchuenche 2011). Using a similar approach, the sensitivity index of $S(t)$ is computed with respect to each of the parameters of model (6) for the three cities (Table 5). In general, E^S, β, σ decrease the student populations as their values increase while I^S increases the student populations as its value increases for all the three cities (Table 5). From Table 5, for the city of Cape Town, the rate σ at which students migrate in search of jobs is the most sensitive parameter followed by the measure of expenditure E^S. This is then followed by the rate β at which students migrate into the business class and finally the measure of income I^S. Thus, the most important parameter that has the influence of changing the student population in Cape Town is the rate at which students migrate in search of jobs. Furthermore, in the city of Durban and Johannesburg, our sensitivity analyses also reveal that the rate σ at which students migrate in search of jobs is the most important

parameter followed by the rate β at which students migrate into the business class. This suggests that in these cities students migrating into the job seeking and business communities should be taken into consideration in constructing more sustainable development policies for the city.

Similar sensitivity analysis to those calculated for $S(t)$ can be carried out to compute the sensitivity indices of $B(t)$, $J(t)$, $W(t)$ and $V(t)$. Even though the sensitivity indices of these remaining solutions are important, we will not present them here to avoid repetition.

Incorporating heterogeneity

Analysis of the homogeneous version of the model reveals that the populations of each of the major South African cities are and will grow fast. However, this analysis does not separate the growth rates of each of the income categories for the city residents. Differentiating these rates should help policymakers to devise better urban development policies. The people of Cape Town, Durban and Johannesburg fall within different income levels (Table 2). Using a case study, and taking this heterogeneity in income levels into account, the population dynamics are explored to help determine the effects of income.

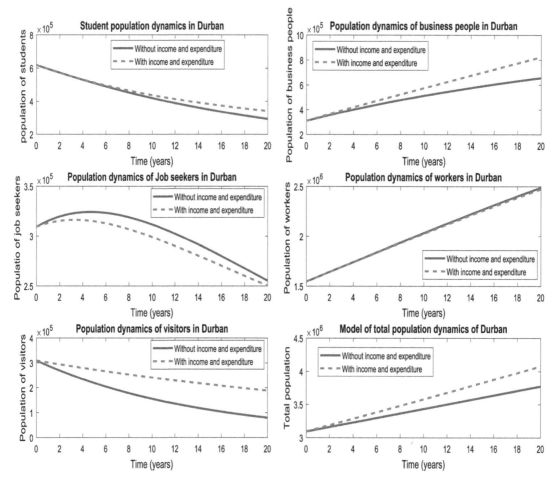

Figure 6: Graphical illustration of the effects of income and expenditure on the population dynamics for various communities (i.e. students, visitors, job seekers, business people, workers) in the city of Durban.

Parameter estimation for the heterogeneous model

To investigate income level population dynamics we estimate the parameter values in the heterogeneous model (1) using Table 2, and our primary and secondary data for the cities of Cape Town, Durban and Johannesburg. The estimated parameter values are presented in Tables 6–12. The coefficients (slopes of the lines) of the linear regressions generated from model findings for each income level in each city are presented in Table 13. From the table, different income levels have different population growth rates. These results show the effect income levels can have on population dynamics.

The estimated internal migration rates of workers w_{ij} and business men b_{ij} shown in Tables 6 and 7 implies that the rate at which workers and business men migrate internally have a large influence on the population dynamics of these cities. These internal migration rates are thus crucial as they help to improve our understanding of city population dynamics with different income levels. Furthermore, from Tables 6–12, there are also significant differences in the estimated parameter values, measures of income and expenditure across different income levels. These differences indicate the presence of heterogeneity in the population dynamics. Therefore, to accurately determine population variations between any two income levels, it is necessary to define an appropriate measure of heterogeneity.

Quantifying heterogeneity

From model (1), there are three main sources of heterogeneity between any two income levels and they are:

(1) Heterogeneity due to differences in income and expenditure.
(2) Heterogeneity due to variation in the movement rates $(\gamma_i, \beta_i, \sigma_i, \delta_i, \alpha_i)$ from one community to another
(3) Heterogeneity due to differences in movement rates of business people b_{ij} and workers w_{ij} across income levels.

Therefore, to determine a measure of heterogeneity between any two income levels, we must consider these three sources of heterogeneity. Using an effective measure of heterogeneity (Collins and Govinder 2014; Robertson, Eisenberg, and Tien 2013; Wilson and Martinez 1997) and defining the distance between two points as Euclidean, a measure of heterogeneity between two income levels is defined as follows:

The first measure of heterogeneity between any two income levels (i and j) due to differences in measures of income and expenditure can be defined as

$$H_{ij}^I = \sqrt{h_{ij}^S + h_{ij}^B + h_{ij}^J + h_{ij}^W + h_{ij}^V} \qquad (8)$$

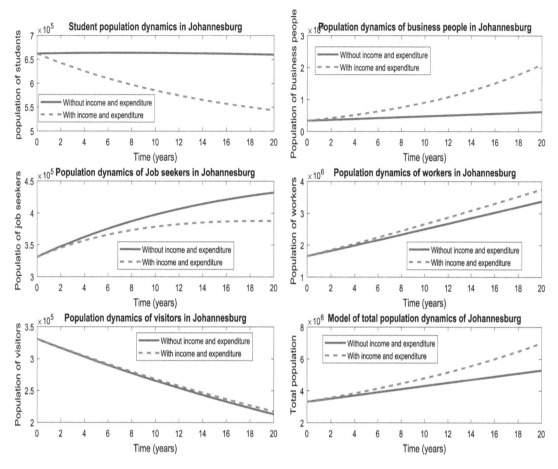

Figure 7: Graphical illustration of the effects of income and expenditure on the population dynamics for various communities (i.e. students, visitors, job seekers, business people, workers) in the city of Johannesburg.

Table 5: Sensitivity indices of $S(t)$ with respect to its parameters for the cities.

Parameter	Symbol	Cape Town	Durban	Johannesburg
Measure of expenditure for S	E^S	−0.1387	−0.0214	−0.1475
Measure of income for S	I^S	0.00079279	0.00088965	0.00089418
Measure of expenditure for B	E^B	0.0000	0.0000	0.0000
Measure of income for B	I^B	0.0000	0.0000	0.0000
Measure of expenditure for J	E^J	0.0000	0.0000	0.0000
Measure of income for J	I^J	0.0000	0.0000	0.0000
Measure of expenditure for W	E^W	0.0000	0.0000	0.0000
Measure of income for W	I^W	0.0000	0.0000	0.0000
Measure of expenditure for V	E^V	0.0000	0.00034761	0.0000
Measure of income for V	I^V	0.0000	−0.1536	0.0000
Migration rate from V to S	γ	0.0000	0.2516	0.0000
Migration rate from S to B	β	−0.0412	−0.3363	−0.1896
Migration rate from S to J	σ	−0.9878	−0.4857	−0.2897
Migration rate from V to J	δ	0.0000	0.0035	0.0000
Migration rate from J to W	α	0.0000	0.0000	0.0000

where $h_{ij}^S = (E_i^S - E_j^S)^2 + (I_i^S - I_j^S)^2$, $h_{ij}^B = (E_i^B - E_j^B)^2 + (I_i^B - I_j^B)^2$, $h_{ij}^J = (E_i^J - E_j^J)^2 + (I_i^J - I_j^J)^2$, $h_{ij}^W = (E_i^W - E_j^W)^2 + (I_i^W - I_j^W)^2$ and $h_{ij}^V = (E_i^V - E_j^V)^2 + (I_i^V - I_j^V)^2$. The quantity h_{ij}^S is the measure of heterogeneity between the student community in income levels i and j due to differences in their income and expenditures. Similarly, h_{ij}^B, h_{ij}^J, h_{ij}^W and h_{ij}^V are heterogeneity measures for the business people, job seekers, workers and visitors respectively.

The second measure of heterogeneity between any two income levels (i and j) which is due to variations in

movement rates (γ_i, β_i, σ_i, δ_i, α_i) can also be defined as

$$H_{ij}^{II} = \sqrt{(\gamma_i - \gamma_j)^2 + (\beta_i - \beta_j)^2 + (\sigma_i - \sigma_j)^2 + (\delta_i - \delta_j)^2 + (\alpha_i - \alpha_j)^2}. \tag{9}$$

The third source of heterogeneity between any two income levels due to differences in movement rates of

Table 6: Estimated internal migration rates b_{ij} for business people in the cities.

Parameter	Symbol	Cape Town	Durban	Johannesburg
Migration rate from B_1 to B_2	b_{12}	0.1716	0.0529	0.1292
Migration rate from B_1 to B_3	b_{13}	0.0979	0.0540	0.1306
Migration rate from B_1 to B_4	b_{14}	0.1361	0.0624	0.1513
Migration rate from B_1 to B_5	b_{15}	0.1303	0.0320	0.1558
Migration rate from B_2 to B_1	b_{21}	0.1280	0.0349	0.0856
Migration rate from B_3 to B_1	b_{31}	0.1916	0.0295	0.1227
Migration rate from B_4 to B_1	b_{41}	0.1361	0.0327	0.0976
Migration rate from B_5 to B_1	b_{51}	0.1408	0.0534	0.1064
Migration rate from B_2 to B_3	b_{23}	0.1234	0.0413	0.0760
Migration rate from B_2 to B_4	b_{24}	0.0824	0.0573	0.0771
Migration rate from B_2 to B_5	b_{25}	0.1244	0.0332	0.1414
Migration rate from B_3 to B_2	b_{32}	0.1292	0.0359	0.1238
Migration rate from B_4 to B_2	b_{42}	0.1734	0.0435	0.1073
Migration rate from B_5 to B_2	b_{52}	0.1032	0.0558	0.1223
Migration rate from B_3 to B_4	b_{34}	0.1530	0.0501	0.1517
Migration rate from B_3 to B_5	b_{35}	0.1810	0.0396	0.1391
Migration rate from B_4 to B_3	b_{43}	0.1559	0.0432	0.1109
Migration rate from B_5 to B_3	b_{53}	0.1277	0.0675	0.1351
Migration rate from B_4 to B_5	b_{45}	0.1737	0.0499	0.1087
Migration rate from B_5 to B_4	b_{54}	0.1126	0.0663	0.1416

Table 7: Estimated internal migration rates w_{ij} for workers in the cities.

Parameter	Symbol	Cape Town	Durban	Johannesburg
Migration rate from W_1 to W_2	w_{12}	0.0891	0.0223	0.0708
Migration rate from W_1 to W_3	w_{13}	0.0803	0.0395	0.0953
Migration rate from W to W_4	w_{14}	0.0248	0.0298	0.1063
Migration rate from W_1 to W_5	w_{15}	0.0452	0.0175	0.1360
Migration rate from W_2 to W_1	w_{21}	0.1610	0.0322	0.0331
Migration rate from W_3 to W_1	w_{31}	0.1131	0.0122	0.0162
Migration rate from W_4 to W_1	w_{41}	0.1621	0.0214	0.0412
Migration rate from W_5 to W_1	w_{51}	0.1316	0.0388	0.0538
Migration rate from W_2 to W_3	w_{23}	0.1125	0.0288	0.1015
Migration rate from W_2 to W_4	w_{24}	0.0823	0.0342	0.1221
Migration rate from W_2 to W_5	w_{25}	0.1024	0.0160	0.0491
Migration rate from W_3 to W_2	w_{32}	0.0709	0.0186	0.0990
Migration rate from W_4 to W_2	w_{42}	0.1319	0.0312	0.1121
Migration rate from W_5 to W_2	w_{52}	0.1248	0.0293	0.1325
Migration rate from W_3 to W_4	w_{34}	0.0581	0.0240	0.1118
Migration rate from W_3 to W_5	w_{35}	0.0706	0.0173	0.0615
Migration rate from W_4 to W_3	w_{43}	0.1655	0.0283	0.1452
Migration rate from W_5 to W_3	w_{53}	0.1848	0.0313	0.0942
Migration rate from W_4 to w_5	w_{45}	0.1223	0.0166	0.0811
Migration rate from W_5 to W_4	w_{54}	0.0616	0.0379	0.1007

Table 8: Estimated measures of income, expenditure and parameter values for income level 1.

Parameter	Symbol	Cape Town	Durban	Johannesburg
Measure of expenditure for S_1	E_1^S	0.0790	0.0342	0.0705
Measure of income for S_1	I_1^S	0.0622	0.0244	0.1215
Measure of expenditure for B_1	E_1^B	0.0693	0.0417	0.1126
Measure of income for B_1	I_1^B	0.1483	0.0582	0.1101
Measure of expenditure for J_1	E_1^J	0.0432	0.0380	0.1408
Measure of income for J_1	I_1^J	0.1471	0.0382	0.0927
Measure of expenditure for W_1	E_1^W	0.0645	0.0220	0.0843
Measure of income for W_1	I_1^W	0.0704	0.0283	0.1078
Measure of expenditure for V_1	E_1^V	0.0120	0.0528	0.0120
Measure of income for V_1	I_1^V	0.0130	0.0527	0.0130
Migration rate from V_1 to S_1	γ_1	0.0100	0.0609	0.0100
Migration rate from S_1 to B_1	β_1	0.0730	0.0299	0.1128
Migration rate from S_1 to J_1	σ_1	0.0673	0.0277	0.0957
Migration rate from V to J_1	δ_1	0.0120	0.0415	0.0120
Migration rate from J_1 to W_1	α_1	0.0907	0.0532	0.0897

Table 9: Estimated measures of income, expenditure and parameter values for income level 2.

Parameter	Symbol	Cape Town	Durban	Johannesburg
Measure of expenditure for S_2	E_2^S	0.1002	0.0340	0.0874
Measure of income for S_2	I_2^S	0.0937	0.0220	0.0462
Measure of expenditure for B_2	E_2^B	0.0916	0.0433	0.0423
Measure of income for B_2	I_2^B	0.1312	0.0505	0.1785
Measure of expenditure for J_2	E_2^J	0.1453	0.0594	0.1263
Measure of income for J_2	I_2^J	0.0857	0.0272	0.0713
Measure of expenditure for W_2	E_2^W	0.0892	0.0286	0.0852
Measure of income for W_2	I_2^W	0.0874	0.0259	0.0730
Measure of expenditure for V_2	E_2^V	0.0120	0.0548	0.0120
Measure of income for V_2	I_2^V	0.0130	0.0491	0.0130
Migration rate from V_2 to S_2	γ_2	0.0100	0.0536	0.0100
Migration rate from S_2 to B_2	β_2	0.0887	0.0309	0.0347
Migration rate from S_2 to J_2	σ_2	0.0955	0.0345	0.0646
Migration rate from V_2 to J_2	δ_2	0.0120	0.0503	0.0120
Migration rate from J_2 to W_2	α_2	0.0725	0.0470	0.0625

Table 10: Estimated measures of income, expenditure and parameter values for income level 3.

Parameter	Symbol	Cape Town	Durban	Johannesburg
Measure of expenditure for S_3	E_3^S	0.1023	0.0236	0.0510
Measure of income for S_3	I_3^S	0.0746	0.0254	0.0445
Measure of expenditure for B_3	E_3^B	0.0714	0.0478	0.0999
Measure of income for B_3	I_3^B	0.1525	0.0359	0.1309
Measure of expenditure for J_3	E_3^J	0.0603	0.0213	0.0415
Measure of income for J_3	I_3^J	0.1510	0.0485	0.0990
Measure of expenditure for W_3	E_3^W	0.1335	0.0223	0.0689
Measure of income for W_3	I_3^W	0.0491	0.0284	0.0749
Measure of expenditure for V_3	E_3^V	0.0120	0.0434	0.0120
Measure of income for V_3	I_3^V	0.0130	0.0530	0.0130
Migration rate from V_3 to S_3	γ_3	0.0100	0.0559	0.0100
Migration rate from S_3 to B_3	β_3	0.0576	0.0298	0.0401
Migration rate from S_3 to J_3	σ_3	0.0763	0.0215	0.0579
Migration rate from V_3 to J_3	δ_3	0.0120	0.0377	0.0120
Migration rate from J_3 to W_3	α_3	0.0905	0.0650	0.0619

business men b_{ij} and workers w_{ij} across the income levels can be defined as

$$H_{ij}^{III} =$$

$$\sqrt{\sum_{k=1}^{5} ((b_{ik} - b_{jk})^2 + (w_{ik} - w_{jk})^2 + (b_{ki} - b_{kj})^2 + (w_{ki} - w_{kj})^2)}.$$

(10)

Based these definitions, the total measure of heterogeneity between any two income levels (i and j) can be estimated as a summation of the above three measures of heterogeneities given by

$$H_{ii} = H_{ii}^{I} + H_{ii}^{II} + H_{ii}^{III}.$$ (11)

Using this definition of heterogeneity together with the estimated parameter values in Tables 6–12, heterogeneity

Table 11: Estimated measures of income, expenditure and parameter values for income level 4.

Parameter	Symbol	Cape Town	Durban	Johannesburg
Measure of expenditure for S_4	E_4^S	0.0919	0.0290	0.0609
Measure of income for S_4	I_4^S	0.1528	0.0250	0.0684
Measure of expenditure for B_4	E_4^B	0.1435	0.0397	0.1314
Measure of income for B_4	I_4^B	0.0971	0.0484	0.0811
Measure of expenditure for J_4	E_4^J	0.1510	0.0484	0.0624
Measure of income for J_4	I_4^J	0.0971	0.0347	0.0984
Measure of expenditure for W_4	E_4^W	0.1118	0.0247	0.0639
Measure of income for W_4	I_4^W	0.1045	0.0241	0.0844
Measure of expenditure for V_4	E_4^V	0.0120	0.0394	0.0120
Measure of income for V_4	I_4^V	0.0130	0.0618	0.0130
Migration rate from V_4 to S_4	γ_4	0.0100	0.0493	0.0100
Migration rate from S_4 to B_4	β_4	0.1245	0.0337	0.0638
Migration rate from S_4 to J_4	σ_4	0.1272	0.0329	0.0789
Migration rate from V_4 to J_4	δ_4	0.0120	0.0509	0.0120
Migration rate from J_4 to W_4	α_4	0.1319	0.0411	0.0571

Table 12: Estimated measures of income, expenditure and parameter values for income level 5.

Parameter	Symbol	Cape Town	Durban	Johannesburg
Measure of expenditure for S_5	E_5^S	0.0783	0.0390	0.0769
Measure of income for S_5	I_5^S	0.0993	0.0237	0.0835
Measure of expenditure for B_5	E_5^B	0.1144	0.0533	0.0793
Measure of income for B_5	I_5^B	0.1193	0.0466	0.1398
Measure of expenditure for J_5	E_5^J	0.1360	0.0536	0.1230
Measure of income for J_5	I_5^J	0.1000	0.0308	0.0972
Measure of expenditure for W_5	E_5^W	0.0582	0.0279	0.1071
Measure of income for W_5	I_5^W	0.1455	0.0340	0.0818
Measure of expenditure for V_5	E_5^V	0.0120	0.0646	0.0120
Measure of income for V_5	I_5^V	0.0130	0.0586	0.0130
Migration rate from V_5 to S_5	γ_5	0.0100	0.0450	0.0100
Migration rate from S_5 to B_5	β_5	0.1046	0.0306	0.0790
Migration rate from S_5 to J_5	σ_5	0.0838	0.0531	0.0720
Migration rate from V_5 to J_5	δ_5	0.0120	0.0387	0.0120
Migration rate from J_5 to W_5	α_5	0.0968	0.0463	0.1008

Table 13: Coefficients of linear regression generated from model fittings for each income level.

Income levels	Cape Town	Durban	Johannesburg
Income level 1	37722	6075.7	9658.6
Income level 2	19406	6075.9	27274
Income level 3	23710	11437	32390
Income level 4	9700.9	8220.2	27275
Income level 5	17241	3931.7	17055

Table 14: Heterogeneity between any two income levels in the city of Cape Town.

H_{ij}	Income level 1	Income level 2	Income level 3	Income level 4	Income level 5
Income level 1	0.0000	0.5927	0.4758	0.6817	0.5548
Income level 2		0.0000	0.5234	0.5405	0.4477
Income level 3			0.0000	0.7042	0.6485
Income level 4				0.0000	0.5209
Income level 5					0.0000

Table 15: Heterogeneity between any two income levels in the city of Durban.

H_{ij}	Income level 1	Income level 2	Income level 3	Income level 4	Income level 5
Income level 1	0.0000	0.1500	0.1577	0.1642	0.1715
Income level 2		0.0000	0.1721	0.1584	0.1577
Income level 3			0.0000	0.1779	0.2220
Income level 4				0.0000	0.2020
Income level 5					0.0000

Table 16: Heterogeneity between any two income levels in the city of Johannesburg.

H_{ij}	Income level 1	Income level 2	Income level 3	Income level 4	Income level 5
Income level 1	0.0000	0.5292	0.5495	0.5173	0.4676
Income level 2		0.0000	0.4371	0.5032	0.4970
Income level 3			0.0000	0.4773	0.5030
Income level 4				0.0000	0.4953
Income level 5					0.0000

between the city income levels are computed (Tables 14–16). Since $H_{ij}=H_{ji}$, the measures of heterogeneity in the lower triangles of the tables are not included to avoid repetition. From Tables 14–16, there are significant differences between any two income levels in the cities particularly in the cities of Cape Town and Johannesburg. Thus, to improve our understanding of city population dynamics using a mathematical model, heterogeneity must be taken into consideration.

Discussion and conclusions

A mathematical model that incorporates heterogeneity in income and expenditure for individuals in a city was formulated to study city population dynamics. Qualitative analyses of the homogeneous version of the model in the absence of income and expenditure showed that the model is globally asymptotically stable and that the total population of a city will not grow unbounded in time. On the other hand, analyses of the homogeneous version of the model in the presence of income and expenditure demonstrated that it is possible for the total population of the city to grow unbounded in time. Further analyses together with these results demonstrate that income and expenditure are major factors that influence city population dynamics.

Case studies of the three major cities in South Africa, Cape Town, Durban and Johannesburg were considered. Using the data from Statistics South Africa (Census 2001 and 2012) and data collected from our respondents in the field, a series of analyses was carried out with our models. The first part of each case study was carried out using the homogeneous version of the model in the presence of income and expenditure and by fitting the model to match the collected data. This also enabled prediction of possible future city population dynamics. Numerical simulations show how income and expenditure can have different but significant effects on the population dynamics of the different population groups (students, workers, visitors, business people, and job seekers). For example, including factors for income and expenditure could have a negative impact on the student populations of Cape Town and Johannesburg but a positive impact on the student population of Durban.

Sensitivity analyses show that the rates at which students migrate in search of jobs and business opportunities are important. This was considered in the second part of each case study using the heterogeneous version of the model. Parameter values were again estimated by fitting the model to real data. Our analyses show that income levels have an important impact on population dynamics. In particular, the rate at which workers and business men migrate internally has a significant influence on city population dynamics. To determine the amount of heterogeneity between any two income levels, a heterogeneity measure was used. Using the estimated parameter values and this measure of heterogeneity, there are significant differences in the dynamics of any two income levels in the South African cities considered. Also, maximum heterogeneity is found at different income levels depending on the cities involved.

In conclusion, for a better description of a city's population dynamics, necessary for effective planning and implementation of policies, it is essential to incorporate heterogeneity in predictive models. The models used here demonstrate the particular importance of heterogeneity in city populations. Also, these dynamics can be unique for a particular city as shown for the South African cities considered here. Thus, policymakers could use results of these types of models in planning. For example, from the results of our models, Johannesburg and Cape Town could prepare for potential increases in business people. These plans could for example incorporate the development of business parks. The same models suggest Durban could prepare for a greater increase in other workers and thus planning could be toward developing more primary industries. Alternatively, the urban planning could be designed to mitigate these population increases if this was the goal of urban management.

The above analyses overall show how city population dynamics can be described by mathematical models that incorporate heterogeneity. In particular, for the three South African cities studied here, heterogeneity in the dynamics of different population groupings are shown to have significant effects on the overall dynamics. Similar models could be developed and used to study the population dynamics of other cities in South African, Africa and worldwide.

Disclosure statement

No potential conflict of interest was reported by the authors.

References

Arieff, A. 2010. *Global Economic Crisis: Impact on Sub-Saharan Africa and Global Policy Responses*. Darby, PA: DIANE Publishing.

Balbo, M., and G. Marconi. 2005. *Governing International Migration in the City of the South*. Global Migration Perspectives #38. Geneva: Global Commission for International Migration.

Brelsford, C., J. Lobo, J. Hand, and L. M. Bettencourt. 2017. "Heterogeneity and Scale of Sustainable Development in Cities." *Proceedings of the National Academy of Sciences* 114: 8963–8968.

Chitnis, N., J. Hyman, and J. Cushing. 2008. "Determining Important Parameters in the Spread of Malaria Through the Sensitivity Analysis of a Mathematical Model." *Bulletin of Mathematical Biology* 70: 1272–1296.

Collins, O. C., and K. S. Govinder. 2014. "Incorporating Heterogeneity Into the Transmission Dynamics of a Waterborne Disease Model." *Journal of Theoretical Biology* 356: 133–143.

Crush, J., and B. Frayne1. 2007. "The Migration and Development Nexus in Southern Africa Introduction." *Development Southern Africa* 24: 1–23.

Deloitte on Africa. 2014, July. The rise and rise of the African middle class. https://www.deloitte.com/view/en in/in/insights-ideas/africa/index.htm.

Epstein, A., Edward M. Bruner, Michael M. Horowitz, Kenneth L. Little, Daniel F. McCall, Phillip Mayer, Horace Miner, Leonard Plotnicov, W. B. Schwab, and William A. Schack. 1967. "Urbanization and Social Change in Africa." *Current Anthropology* 8 (4): 275–295.

Goldstein, S. 1990. "Urbanization in China, 1982–87: Effects of Migration and Reclassification." *Population and Development Review* 16: 673–701.

Gounden, A., and T. Nkhumeleni. 2013. "The rise and rise of the Africa middle class." Vol. 1.

Harris, J. R., and M. P. Todaro. 1970. "Migration, Unemployment and Development: A Two Sector Analysis." *The American Economic Review* 60 (1): 126–142.

Hatton, T. J., and M. Tani. 2005. "Immigration and Inter-Regional Mobility in the UK, 1982–2000." *The Economic Journal* 115: F342–FF35.

Hove-Musekwa, S., F. Nyabadza, C. Chiyaka, P. Das, A. Tripathi, and Z. Mukandavire. 2011. "Modelling and Analysis of the Effects of Malnutrition in the Spread of Cholera." *Mathematical and Computer Modelling* 53: 1583–1595.

Kihato, C. 2018. NEPAD, the city and the migrant: implications for urban Governance. Migration Policy Brief (12), Southern African Migration Project.

Kok, P., and M. Collinson. 2006. *Migration and Urbanisation in South Africa*. Pretoria: Statistics South Africa.

Montgomery, M. R. 2008. "The Urban Transformation of the Developing World." *Science* 319: 761–764.

Mwasa, A., and J. Tchuenche. 2011. "Mathematical Analysis of a Cholera Model with Public Health Interventions." *BioSystems* 105: 190–200.

Odularu, G. O. 2014, June. "Crude oil and the Nigerian economic performance." http://www.ogbus.ru/eng/authors/Odularo/Odularo1.pdf.

Onwe, O. J. 2013. "Role of the Informal Sector in Development of the Nigerian Economy: Output and Employment Approach." *Journal of Economics and Development Studies* 1 (1): 60–74.

Pearl, R., and L. Reed. 1920. "On the Rate of Growth of the Population of United States Since 1790 and its Mathematical Representation." *Proceedings of the National Academy of Sciences of the USA* 6: 275–288.

Rapanos, V. T. 2005. "Minimum Wage and Income Distribution in the Harris-Todaro Model." *Journal of Economic Development* 30: 1–14.

Rees, D., J. Murray, G. Nelson, and P. Sonnenberg. 2010. "Oscillating Migration and the Epidemics of Silicosis, Tuberculosis, and HIV Infection in South African Gold Miners." *American Journal of Industrial Medicine* 53: 398–404.

Robertson, S. L., M. C. Eisenberg, and J. H. Tien. 2013. "Heterogeneity in Multiple Transmission Pathways: Modelling the Spread of Cholera and Other Waterborne Disease in Networks with a Common Water Source." *Journal of Biological Dynamics* 7: 254–275.

Solecki, W., K. C. Seto, and P. J. Marcotullio. 2013. "It's Time for an Urbanization Science." *Environment: Science and Policy for Sustainable Development* 55: 12–17.

Statistics South Africa census 2001 and census 2011. 2014, May. http://beta2.statssa.gov.za/.

Todaro, M. 1997. "Urbanisation, unemployment and urban migration in Africa theory and policy." *Working paper* Vol. 104.

Tsoularis, A. 2001. "Analysis of Logistic Growth Models." *Research Letters in the Information and Mathematical Sciences* 2: 23–46.

UN-Habitat. 2012. State of the World's Cities 2012/2013, Prosperity of Cities. https://sustainabledevelopment.un.org/content/documents/745habitat.pdf.

Van Hear, N., O. Bakewell, and K. Long. 2018. "Push-pull Plus: Reconsidering the Drivers of Migration." *Journal of Ethnic and Migration Studies* 44: 927–944.

Vearey, J., I. Palmary, L. Thomas, L. Nunez, and S. Drimie. 2010. "Urban Health in Johannesburg: the Importance of Place in Understanding Intra-Urban Inequalities in a Context of Migration and HIV." *Health & place* 16: 694–702.

Verhulst, P. 1838. "Notice sur la loi que la population suit dans son accroissement." *Correspondance mathmatique et physique* 10: 113–121.

Wilson, D., and T. R. Martinez. 1997. "Improved Heterogeneous Distance Functions." *Journal of Artificial Intelligence Research* 6: 1–34.

Zenou, Y., and T. Smith. 1995. "Efficiency Wages, Involuntary Unemployment and Urban Spatial Structure." *Regional Science and Urban Economics* 25: 821–845.

Mathematical model showing how socioeconomic dynamics in African cities could widen or reduce inequality

Obiora Cornelius Collins, Thokozani Silas Simelane and Kevin Jan Duffy

Cities are important forces of national socioeconomic development. Individuals in cities often belong to different socioeconomic statuses depending on their levels of income, education and nature of occupation. Income, employment and education opportunities are among the main attractions of most cities. In this study, we investigate the impact of socioeconomic status on city attractiveness for the African cities of Windhoek, Harare, Lusaka, Kinshasa and Nairobi. The socioeconomic status of samples of individuals in these cities are used to formulate a mathematical model that describes the city population dynamics. Using income as a measure, qualitative analyses of the model together with numerical simulations using survey data show how competitive relationships among the various socioeconomic status groups could widen inequality over time. Alternatively, synergetic relationships among the various socioeconomic groups could reduce this inequality. These results point to urban planning that encourages synergism between the different income classes with the aim of reducing inequality.

Introduction

Many African cities face challenges in terms of the quality of life they offer their citizens. Despite widespread agreement on the need for improvement, effective programmes are seldom successfully implemented in most cities in Africa (Turok and Watson 2001). In the absence of effective urban policies many African cities are less likely to be inclusive, safe, resilient and sustainable as desired by Sustainable Development Goal Number 11 (Radoslav et al. 2012). In addition, improving financial security and job satisfaction in most cities in Africa are less likely. Without these attributes African cities are not expected to be attractive to move to and settle in. This study considers the importance of income and education on the population dynamics of cities. Understanding these processes can assist in formulating effective policies for improving the quality of life provided by these cities.

Cities naturally originate from a complex history of economic, political and spatial factors and are thus multifaceted with no one description. However, there are recent empirical and theoretical studies that have shown that most cities have properties that are, to some extent, quantitatively predictable (Bettencourt 2013). These conclusions are confirmed from different fields of study such as economics (Glaeser et al. 2012), geography (Batty 2008) and complex systems analyses (Bettencourt et al. 2007). In essence, these studies show how cities develop by attractive forces, and their opposite, based on socioeconomic benefits verses costs. While this approach has its limitations, it can help with understanding certain temporal population growth scenarios and this is the direction taken in this paper.

Cities provide important forces of national and regional socioeconomic development (Ratcliffe 2010). According to Snieska and Zykiene (2015), the level of investment in a city is influenced by accessibility of a skilled workforce and its costs, the price of resources and competition in the market. Employment and education are factors that can improve the chances of generating income and thereby enhance the socioeconomic status for individuals and their dependents. Thus, understanding how income and education contribute to the transformation of a city is important for urban management aligned towards improving city quality.

Socioeconomic status depends largely on level of occupation, income and education (Musterd et al. 2017). Socioeconomic status can be broken into levels (for example, low, middle and high levels) and individuals can be placed into these categories using any or all of the three variables (income, education and occupation). Individuals are likely to be attracted to a city where there are more chances of an improved socioeconomic status. There are however also internal factors to any city's dynamics. In particular, socioeconomic segregation has been shown to have a dominant influence on how cities evolve with evidence in Europe of an increasing divide between these status groups (Musterd et al. 2017). Thus, knowledge of socioeconomic status and its dynamics are important for urban planning (Kabudul et al. 2017).

A significant amount of research has been done on the shape and structure of urban growth patterns (Batty 2008; Bettencourt 2013; Li et al. 2017; Louf and Barthelemy 2013; Simini et al. 2012). These studies generally consider spatial scales rather than the temporal aspects; yet, to understand the overall dynamics of a city, it is also necessary to consider these temporal aspect of population growth and the drivers of such growth. This approach had been used for city population migration by Weidlich and Haag (2012). In this paper, we consider temporal population growth for African cities using the approach of Bettencourt et al. (2007). This approach uses population dynamic models with migration driven by people looking for economic opportunities.

The impact of socioeconomic status on the population dynamics of certain African Cities is considered. A field survey was conducted to determine the socioeconomic status of individuals for 10 African cities, namely: Port

Elizabeth, Cape Town, Durban, Johannesburg, Pretoria, Windhoek, Harare, Lusaka, Kinshasa and Nairobi. The results of this survey were very different for cities within and outside of South Africa. In South Africa, there is a variety of socioeconomic status distributions and this is considered in a previous paper (Duffy, Simelane, and Collins 2018). For all the cities outside South Africa, low socioeconomic status in terms of income has the greatest proportion (above 80%) of the population, while a very small proportion has a middle or high socioeconomic status. A mathematical model is used here to determine the impact of this status on the population dynamics of the city.

The mathematical model developed here is designed to take socioeconomic status into consideration in the city population dynamics. While drivers of migration can be due to economic, social, political, environmental or demographic factors (Black et al. 2011), income and education are generally primary drivers of city transformation (Duffy, Simelane, and Collins 2018). However, as explained below, there can be differences in the distribution of income and education for a particular African city. However, a number of studies have found that migration is primarily influenced by the need to find employment (Black et al. 2011; Castles 2013; Van Hear, Bakewell, and Long 2018) and employment has been used alone to successfully model mobility and migration patterns (Simini et al. 2012), so this factor was chosen to represent the indicator of socioeconomic status in our models. Considering the existence of differences in income levels across cities, the possible determinants for how income influences city dynamics over time are considered through simulations of the model. Various scenarios through these simulations could be used to assist in determining city planning. In particular, we investigate the effects of positive or negative interactions between the socioeconomic status groups and show how these can either increase or decrease the income divide between these groups. Thus, we predict that if countering measures are not implemented there will be an increased divide between the economic classes of African cities as seen in European cities (Musterd, 2017).

Materials and method
Data collection methodology
A field survey to determine the socioeconomic status of individuals within a chosen group of African cities was conducted. Two socioeconomic variables, income and education, were considered as measures of classification. The classification of these factors is described below. Data was collected using questionnaires. In each city, a maximum of 200 and a minimum of 100 participants were interviewed, with targeted ages of respondents ranging between 20 and 50 years. A number of variables were considered including the income and educational levels of respondents. The cities considered here are Kinshasa (DRC), Lusaka (Zambia), Harare (Zimbabwe), Nairobi (Kenya) and Windhoek (Namibia).

Income
Classification of individuals according to socioeconomic status based on their monthly income is based on the mean monthly minimum wages for seven African countries

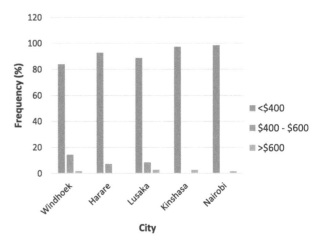

Figure 1: The percentage frequencies of socioeconomic classes of individuals based on income for African cities.

($457) and the distribution of these wages (Bhorat, Kanbur, and Stanwix 2017) and are set as follows: a low socioeconomic status (N_1) are people whose monthly income is less than $400, an average socioeconomic status (N_2) are people whose monthly income is between $400 to $600 and a high socioeconomic status (N_3) are people whose monthly income is above $600. The results of surveys conducted in Windhoek, Harare, Lusaka, Kinshasa and Nairobi are presented in the Figure 1. As can be noted, low socioeconomic status has the greatest proportion (above 80%) of the population in these cities, with a very small proportion in the middle and high socioeconomic status groups.

Education
Another socioeconomic variable considered in the classification of individuals is education. For this, we categorized individuals as follows: a low socioeconomic status (N_1) are people with a matric certificate or below, an average socioeconomic status (N_2) are people with a diploma or a degree qualification and a high socioeconomic status (N_3) are people with a postgraduate qualification. The results using education as a measure of classification of socioeconomic status are presented in Figure 2.

There is a significant difference between results obtained using income as a measure of socioeconomic status and results obtained using education. For the rest

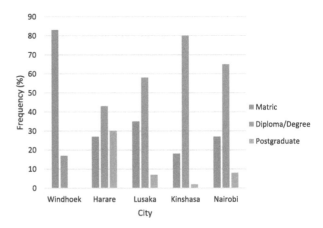

Figure 2: The percentage frequencies of socioeconomic classes of individuals based on education for African cities.

of the analyses, income is chosen as the measure of classification of socioeconomic statuses in the cities. The reason for this choice was that while education is important, we decided that income was more likely to influence people in their decisions on permanent residence. This decision is backed up by a number of studies that found that migration is primarily influenced by the need to find employment (Black et al. 2011; Castles 2013; Van Hear, Bakewell, and Long 2018). However, the methodology could also be used for other groupings of socioeconomic status such as using educational status.

Mathematical model for city dynamics

For the model developed here the total city population denoted by N is subdivided in three categories N_1, N_2, N_3 representing the low, middle and high socioeconomic population classes respectively. The growth of the population of each of these socioeconomic status groups is modelled using logistic population growth (Collins and Duffy 2016). People changing their socioeconomic status class from N_i to another class N_j do so at a rate δ_{ij}. The nature of interactions of the various socioeconomic classes is important in the population dynamics of the city. This type of interaction can be synergetic or not. The nature of interactions can be complex as a result of complex human behaviour. We illustrate this with a very simple example. Individuals with higher socioeconomic status are often the employers of people with lower socioeconomic status. For these cases, when there is a mutual relationship between an employer and an employee, both can benefit synergistically from each other. On the other hand, when they disagree on conditions of service, for example, or wage negotiation, the interaction is uncooperative. To incorporate these interactions into the model, we assume that b_{ij} is a measure of synergetic effects of N_j on N_i if the value is positive. These interactions are assumed to have the opposite effect if b_{ij} is negative, that is when uncooperative effects dominate. If b_{ij} is zero these effects are balanced. Based on these assumptions and formulations, we obtain the model:

$$
\frac{dN_1}{dt} = N_1 r_1 \left(1 - \frac{N_1}{K_1} - b_{12}\frac{N_2}{K_1} - b_{13}\frac{N_3}{K_1}\right)
$$
$$
+ \delta_{21}N_2 - \delta_{12}N_1
$$
$$
\frac{dN_2}{dt} = N_2 r_2 \left(1 - \frac{N_2}{K_2} - b_{21}\frac{N_1}{K_2} - b_{23}\frac{N_3}{K_2}\right)
$$
$$
+ \sum_{i=1}^{3}\delta_{i2}N_i - \sum_{i=1}^{3}\delta_{2i}N_2
$$
$$
\frac{dN_3}{dt} = N_3 r_3 \left(1 - \frac{N_3}{K_3} - b_{31}\frac{N_1}{K_3} - b_{32}\frac{N_2}{K_3}\right)
$$
$$
+ \delta_{23}N_2 - \delta_{32}N_3, \tag{1}
$$

with initial conditions: $N_1(0) > 0$, $N_2(0) > 0$, $N_3(0) > 0$.

Mathematical analyses of the model

The mathematical analyses of model (1) are used to gain insight into the population dynamics of each socioeconomic

class for a city. The long- and short-term dynamics of a dynamical system model can be described by the stability about its equilibrium points (Liao and Wang 2011) and so stability analyses are used to investigate the dynamics of model (1). The following cases are considered:

Case 1. The first case is a situation when there are no changes in socioeconomic status ($\delta_{ij} = 0$) and there is neither synergetic nor the opposite ($b_{ij} = 0$). Then model (1) has the following equilibrium points:

$$
E_1 = (0, 0, 0), E_2 = (K_1, 0, 0), E_3 = (0, K_2, 0),
$$
$$
E_4 = (0, 0, K_3), E_5 = (K_1, K_2, 0), E_6 = (K_1, 0, K_3),
$$
$$
E_7 = (0, K_2, K_3), E_8 = (K_1, K_2, K_3). \tag{2}
$$

Theorem 1. *The equilibrium points:* E_1, E_2, E_3, E_4, E_5, E_6, E_7 *are unstable while the positive equilibrium point* E_8 *is globally asymptotically stable.*

The instability of the equilibrium points: E_1, E_2, E_3, E_4, E_5, E_6, E_7 can be established by showing the Jacobian of model (1) evaluated at each of them has an eigenvalue with a positive real part. Take for example the trivial equilibrium point E_1. The Jacobian of model (1) evaluated at the trivial equilibrium point E_1 has the following positive eigenvalues: $\lambda_1 = r_1 > 0, \lambda_2 = r_2 > 0$ and $\lambda_3 = r_3 > 0$. Thus, the trivial equilibrium point E_1 is unstable. The implication of this result is that based on our model formulations, it will be difficult for the population of any African city to go extinct. Similarly, for E_2 we have that the Jacobian of model (1) evaluated at the equilibrium point E_2 has the following eigenvalues: $\lambda_1 = -r_1 < 0, \lambda_2 = r_2 > 0$ and $\lambda_3 = r_3 > 0$. Thus, E_2 is unstable. The implication of this result is that based on our model formulations, it will be difficult for only the lower socioeconomic status group to survive in any African city that comprises all three socioeconomic status groups. By similar arguments we can show that the equilibrium points: E_3, E_4, E_5, E_6, E_7 are unstable. The mathematical implications of these results are that it will be difficult for a city that comprises all three socioeconomic status groups to survive with only one or two socioeconomic status groups. The asymptotic stability of the positive equilibrium point E_8 can be established by solving the model (1) analytically and showing that the solution of the model converges to the equilibrium point E_8 as time tends to infinity. The exact solution of model (1) is given by

$$
(N_1(t), N_2(t), N_3(t)) = \left(\frac{C_1 K_1 e^{r_1 t}}{K_1 + C_1 e^{r_1 t}}, \frac{C_2 K_2 e^{r_2 t}}{K_2 + C_2 e^{r_2 t}}, \frac{C_3 K_3 e^{r_3 t}}{K_3 + C_3 e^{r_3 t}}\right), \tag{3}
$$

where $C_1 = \frac{K_1 N_1(0)}{K_1 - N_1(0)}, C_2 = \frac{K_2 N_2(0)}{K_2 - N_2(0)}$ and $C_3 = \frac{K_3 N_3(0)}{K_3 - N_3(0)}$.

By elementary algebraic manipulation, $(N_1(t), N_2(t), N_3(t)) \to (K_1, K_2, K_3)$ as $t \to \infty$. This proves that the equilibrium point E_8 is globally asymptotically stable. Mathematically, this shows that the three socioeconomic status groups can coexist (survive) for a long period of time if no changes in socioeconomic status are possible and there are no synergetic interactions.

Table 1: Parameter values and their sources for some African cities.

Parameters	Meaning	Windhoek	Harare	Lusaka	Kinshasa	Nairobi	Source
r_i	Growth rate	2.13%	2.32%	3.01%	3.09%	2.52%	WPR 2018
$N(0)$	Initial population size	322000	1560000	1700000	7800000	3500000	WPR 2018
$N_i(0)$	Initial population size for SEC i	$p_i N(0)$	$p_i N(0)$	$p_i N(0)$	$p_i N(0)$	$p_i N(0)$	Estimated
K_i	Carrying capacity	$10 p_i N(0)$	$10 p_i N(0)$	$10 p_i N(0)$	$10 p_i N(0)$	$10 p_i N(0)$	Estimated

Note: p_i is the proportion of individual in N_i (see Figure 1(b) for the cities of Windhoek, Harare, Lusaka, Kinshasa and Nairobi)

Case 2. The second case is a situation when there are no changes in socioeconomic status (i.e., $\delta_{ij} = 0$) but $b_{ij} \neq 0$. Then model (1) has the following equilibrium points:

$$E_1 = (0, 0, 0), E_2 = (K_1, 0, 0), E_3 = (0, K_2, 0),$$
$$E_4 = (0, 0, K_3),$$

$$E_5 = \left(\frac{K_1 - b_{12}K_2}{1 - b_{12}\, b_{21}}, \frac{K_2 - b_{21}K_1}{1 - b_{12}\, b_{21}}, 0 \right),$$

$$E_6 = \left(\frac{K_1 - b_{13}K_3}{1 - b_{13}\, b_{31}}, 0, \frac{K_3 - b_{31}K_1}{1 - b_{13}\, b_{31}} \right),$$

$$E_7 = \left(0, \frac{K_2 - b_{23}K_3}{1 - b_{23}\, b_{32}}, \frac{K_3 - b_{32}K_2}{1 - b_{23}\, b_{32}} \right),$$

$$E_8 = \left(\frac{K_1}{2}, \frac{K_1}{2}, \frac{K_1}{2} \right). \tag{4}$$

Note the positive equilibrium E_8 is obtained for a special case when $K_1 = K_2 = K_3$ and $b_{ij} = (1/2)$.

Theorem 2. *The equilibrium points: E_1, E_2, E_3, E_4, E_5, E_6, E_7 are unstable while the positive equilibrium point E_8 is stable.*

The proof of Theorem 2 can be established using a similar approach as in the proof of Theorem 1.

The analytical solution of the general case of model (1) in the presence of changes in socioeconomic status (i.e., $\delta_{ij} \neq 0$) and a synergetic or uncooperative relationship between classes (i.e., $b_{ij} \neq 0$) is too complex to solve analytically because the differential equation is non-linear. Thus, to gain insight into the dynamics of the model, we consider numerical solutions.

Numerical results

To investigate the dynamics of model (1) numerically for Windhoek, Harare, Lusaka, Kinshasa and Nairobi, we first used demographic data of each of the cities for the period 1995–2018 from the World Population review 2018 (see Table 1). The remaining parameter values are estimated by using the model fit to the population dynamic data for each city for the same period. The model fittings were carried out using the built-in MATLAB least-squares fitting routine Fmincon in the optimization tool box of the software. The results of the fitting for these cities are given in Figure 3 and the associated parameter estimates are given in Table 2. These results illustrate

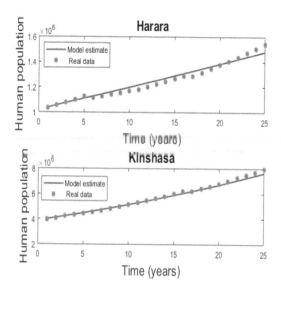

Figure 3. Model fitting of the population dynamics of some African cities (Windhoek, Harare, Lusaka, Kinshasa and Nairobi) from 1995 to 2018. The bold lines represent the model fit and the stars represent the real data.

Table 2: Estimated parameter values from model fitting of African cities.

Parameters	Windhoek	Harare	Lusaka	Kinshasa	Nairobi
b_{12}	0.3545	0.2518	0.2041	0.1011	0.1306
b_{13}	0.3290	0.2066	0.1976	0.1013	0.1452
b_{21}	0.2822	0.8369	0.2320	0.1000	0.1034
b_{23}	0.3284	0.1916	0.1924	0.1001	0.1584
b_{31}	0.2283	0.8613	0.2115	0.1000	0.1180
b_{32}	0.2643	0.1952	0.1958	0.1001	0.1659
δ_{12}	0.2089	0.1617	0.2028	0.1000	0.1136
δ_{21}	0.7326	0.8994	0.7522	0.9000	0.9000
δ_{23}	0.2012	0.1092	0.1908	0.1267	0.1000
δ_{32}	0.7594	0.8994	0.7691	0.1004	0.2041

that our model can be used to study and predict socioeconomic (population) dynamics for each of these cities.

Next, we investigated the impact of socioeconomic status on the population dynamics of these cities. With an uncooperative relationship among the socioeconomic groups, numerical simulations of the population dynamics for the five cities are presented in Figure 4. From the figure, the population of the low socioeconomic status N_1 dominates in all these cities and continues to increase faster than for the other socioeconomic groups. In general, socioeconomic dynamics of all these cities combined over a 100-year period result in a similar trend (Figure 5).

With a synergetic relationship among the socioeconomic groups, numerical simulations of the population dynamics for the five cities are presented in Figure 6. From the figure, the low socioeconomic status group N_1

dominates in Windhoek, Lusaka, Kinshasa and Nairobi but not in Harare. Further analyses (Figure 7) demonstrate that for a synergetic relationship, the gap in population size among the various socioeconomic groups is not as extreme when compared with the case of uncooperative relationships between status groups.

Discussion

Like most world cities (Gaigné et al. 2017), African cities are made up of individuals who belong to distinct socioeconomic status groups. In the African cities studied here, these socioeconomic differences have a bearing on the city dynamics. In particular, interactions among the socioeconomic groups determine the overall population dynamics of the city. The relationships among the socioeconomic groups can be either synergetic or uncooperative. The effects on the dynamics, as dependent on internal

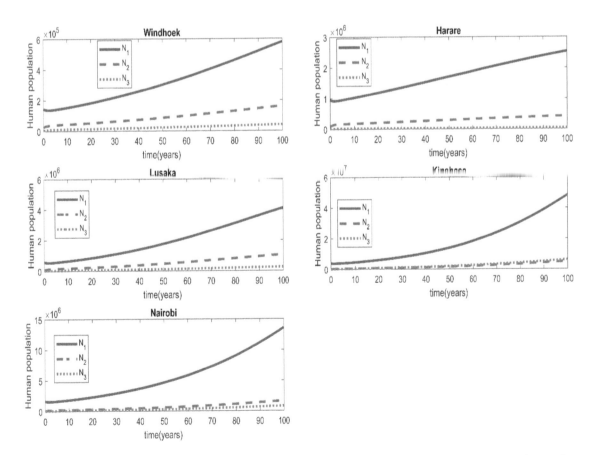

Figure 4: Possible population dynamics of cities (Windhoek, Harare, Lusaka, Kinshasa and Nairobi) for a scenario where various socioeconomic classes in the cities have an uncooperative relationship.

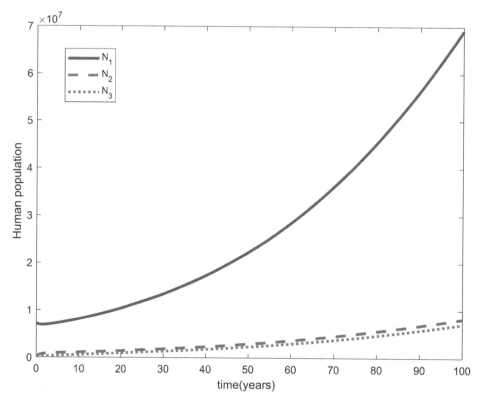

Figure 5: Illustration of the possible socioeconomic dynamics of the cities (Windhoek, Harare, Lusaka, Kinshasa and Nairobi) for a scenario where various socioeconomic classes in the city have an uncooperative relationship. Socioeconomic dynamics of for all these cities combined over 100 years.

population growth parameters and the possible relationships between the classes, are explored here using a mathematical model.

We first considered the current state of groupings of socioeconomic status in five African cities. Both income and education were considered as measures of

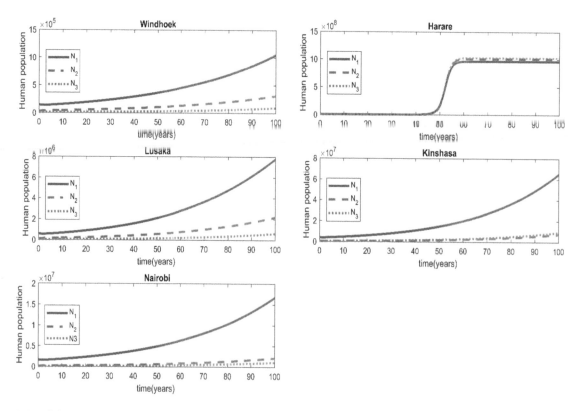

Figure 6: Possible population dynamics of the cities Windhoek, Harare, Lusaka, Kinshasa and Nairobi for a scenario where the socioeconomic classes in the city have synergistic relationships.

socioeconomic status. From the point of view of income, all the cities have a high proportion of people of low socioeconomic status. There are significant differences between these results when education rather than income is used as a measure of socioeconomic status. When education is used as the measure of class, most city dwellers are found in the medium socioeconomic status group. A possible explanation for the difference could be the high levels of unemployment due to economic conditions regardless of education. For instance, in Kinshasa about 80% of youth from ages 10–15 were enrolled at school in the 1990s. This is relatively high for a city that had poor economic conditions over the same period (Mabika and Shapiro 2012). Also, studies have shown that education alone does not necessarily reduce differences between economic groups (Marginson 2016).

Next a mathematical model was developed to explore the dynamics further. For this model, income was used as the primary measure of socioeconomic status. Overall dynamics from the model show that for the African cities outside of South Africa the differences between the low, medium and higher economic classes generally increase, with the low class increasing proportionally the most. These results contrast with South African cities where a similar model showed how the middle-income class could grow proportionally, compared with the other classes (Duffy, Simelane, and Collins 2018).

Analyses based on the model used here and using the survey data show how separation between socioeconomic groups could increase as found in many cities of the world, for example in Europe (Musterd et al. 2017; Tammaru et al. 2015). Our results show that uncooperative relationships between status groups can increase this separation between higher income earners and the poorer ones. Thus, uncooperative relationships create less conducive environments for healthy city balance. The model then shows how synergistic relationships among the socioeconomic status groups can reduce these differences (as compared to the uncooperative examples). This possibility is very important as it shows how differences in proportions of higher and poorer income earners can be reduced if measures are put in place by planners and managers to enable and encourage better cohesion between socioeconomic groups. In particular, for the city of Harare, over time the proportions of the different classes equalize with synergetic effects between classes. This possibility is also very interesting. It appears to suggest that some inherent aspect of the existing Harare population dynamic could lead to a very much more equal society, even though the existing economic base is very low, if these types of synergetic planning methods could be established.

These simulation results exemplify other findings in which the divide in economic status is increasing (Bradbury 2016) and it is important to establish urban policies that work against these trends, such as policies that support synergistic effects between groups, for example, enabling different income scales to exist in adjacent urban areas. Often urban agglomeration benefits mainly the wealthier segments in pockets of a city which can then lead to an

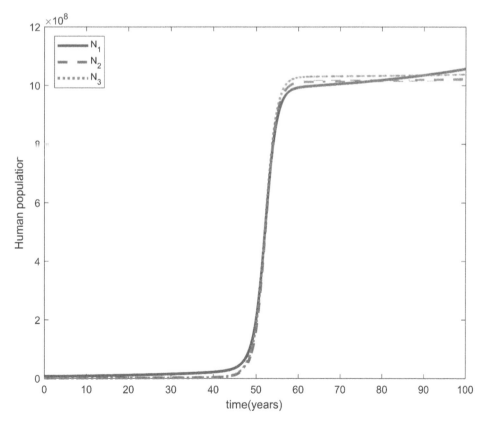

Figure 7: Illustration of the possible socioeconomic dynamics of the combined cities Windhoek, Harare, Lusaka, Kinshasa and Nairobi for a scenario where the socioeconomic classes in the city have synergistic relationships. That is, the socioeconomic dynamics of all these cities combined over 100 years.

increase in the socioeconomic divide (Bettencourt and Lobo 2016; Sarkar et al. 2018). Transport, education, housing and other amenities spread more evenly across a city should also ameliorate these effects.

While models like the one presented here are very simple and leave out many of the real dynamics, they can be used as a first step in understanding the system. Already the model developed here illustrates how important some of the dynamics can be. Also, the model can be extended to probe more complex scenarios of city transformation by being incorporated into larger system dynamic models. In this way, they can be developed to be more representative of real-world situations. In a subtle way, the relationships presented illustrate the influence of income on city dynamics and transformation, which come into being through population dynamics as people move from one socioeconomic status to another. The model presented here already points to potential planning that could reduce inequality in African cities.

Disclosure statement

No potential conflict of interest was reported by the authors.

References

Batty, Michael. 2008. "The Size, Scale, and Shape of Cities." *Science* 319: 769–771.

Bettencourt, Luís M.A. 2013. "The Origins of Scaling in Cities." *Science* 340: 1438–1441.

Bettencourt, Luís M.A., and José Lobo. 2016. "Urban Scaling in Europe." *Journal of The Royal Society Interface* 13: 20160005.

Bettencourt, Luís M.A., José Lobo, Dirk Helbing, Christian Kühnert, and Geoffrey B. West. 2007. "Growth, Innovation, Scaling, and the Pace of Life in Cities." *Proceedings of the National Academy of Sciences* 104 (17): 7301–7306.

Bhorat, Haroon, Ravi Kanbur, and Benjamin Stanwix. 2017. "Minimum Wages in Sub-Saharan Africa: A Primer." *The World Bank Research Observer* 32 (1): 21–74.

Black, Richard, W. Neil Adger, Nigel W. Arnell, Stefan Dercon, Andrew Geddes, and David Thomas. 2011. "The Effect of Environmental Change on Human Migration." *Global Environmental Change* 21: S3–S11.

Bradbury, Bruce. 2016. "Spatial Inequality of Australian Men's Incomes, 1991 to 2011." Australian labour market research conference, Canberra.

Castles, Stephen. 2013. "The Forces Driving Global Migration." *Journal of Intercultural Studies* 34 (2): 122–140.

Collins, Obiora Cornelius, and Kevin Jan Duffy. 2016. "Consumption Threshold Used to Investigate Stability and Ecological Dominance in Consumer-Resource Dynamics." *Ecological Modelling* 319: 155–162.

Duffy, Kevin J., Thokozani S. Simelane, and Obiora C. Collins. 2018. "Income as a Primary Driver of South African Inner City Migration." *Theoretical and Empirical Researches in Urban Management* 13 (3): 25–36.

Gaigné, Carl, Hans R. A. Koster, Fabien Moizeau, and Jacques-François Thisse. 2017. *Amenities and the Social Structure of Cities*. Retrieved from https://pet2017paris2.sciencesconf.org/148299/document.

Glaeser, Edward, Joshua Gottlieb, and Joseph Gyourko. 2012. "Can Cheap Credit Explain the Housing Boom?" In *Housing and the Financial Crisis*, edited by Edward L. Glaeser and Todd Sinai, 301–359. Chicago: University of Chicago Press.

Kabudula, Chodziwadziwa W., Brian Houle, Mark A. Collinson, Kathleen Kahn, Stephen Tollman, and Samuel Clark. 2017. "Assessing Changes in Household Socioeconomic Status in Rural South Africa, 2001–2013: A Distributional Analysis Using Household Asset Indicators." *Social Indicators Research* 133 (3): 1047–1073.

Li, Ruiqi, Lei Dong, Jiang Zhang, Xinran Wang, Wen-Xu Wang, Zengru Di, and H. Eugene Stanley. 2017. "Simple Spatial Scaling Rules Behind Complex Cities." *Nature Communications* 8 (1): 1841.

Liao, Shu, and Jin Wang. 2011. "Stability Analysis and Application of a Mathematical Cholera Model." *Mathematical Biosciences and Engineering* 8 (3): 733–752.

Louf, Rémi, and Marc Barthelemy. 2013. "Modeling the Polycentric Transition of Cities." *Physical Review Letters* 111 (19): 198702.

Mabika, Crispin Mabika, and David Shapiro. 2012. "School Enrollment in the Democratic Republic of the Congo: Family Economic Well-Being, Gender and Place of Residence." *African Population Studies* 26 (2). https://tspace.library.utoronto.ca/bitstream/1807/49299/1/ep12011.pdf

Marginson, Simon. 2016. "The Worldwide Trend to High Participation Higher Education: Dynamics of Social Stratification in Inclusive Systems." *Higher Education* 72 (4): 413–434.

Musterd, Sako, Szymom Marcińczak, Maarten Van Ham, and Tjit Tammaru. 2017. "Socioeconomic Segregation in European Capital Cities. Increasing Separation Between Poor and Rich." *Urban Geography* 38: 1062–1083.

Radoslav, Radu, Marius Stelian Găman, Tudor Morar, Ştefana Bădescu, and Ana Maria Branea. 2012. "Sustainable Urban Development Through the Empowering of Local Communities." Sustainable development-policy and urban development-tourism, life science, management and environment. http://cdn.intechopen.com/pdfs/29222.pdf.

Ratcliffe, J. 2010. "Competitive Cities: Five Keys to Success." Accessed 20 January 2010. http://www. chforum. org/library/com pet_cities. shtml.

Sarkar, Somwrita, Peter Phibbs, Roderick Simpson, and Sachin Wasnik. 2018. "The Scaling of Income Distribution in Australia: Possible Relationships Between Urban Allometry, City Size, and Economic Inequality." *Environment and Planning B: Urban Analytics and City Science* 45 (4): 603–622.

Simini, Filippo, Marta C. González, Amos Maritan, and Albert-László Barabási. 2012. "A Universal Model for Mobility and Migration Patterns." *Nature* 484 (7392): 96.

Snieska, Vytautas, and Ineta Zykiene. 2015. "City Attractiveness for Investment. Characteristics and Underlying Factors. *Procedia-Social and Behavioral Sciences* 213: 48 54.

Tammaru, Tiit, Szymon Marcińczak, Maarten Van Ham, and Sako Musterd. 2015. "A Multi-Factor Approach to Understanding Socio-Economic Segregation in European Capital Cities." In *Socio-Economic Segregation in European Capital Cities*, 25–53. Abingdon, UK: Routledge.

Turok, Ivan, and Vanessa Watson. 2001. "Divergent Development in South African Cities: Strategic Challenges Facing Cape Town." *Urban Forum* 12 (2): 119–138. Springer-Verlag.

Van Hear, Nicholas, Oliver Bakewell, and Katy Long. 2018. "Push-Pull Plus: Reconsidering the Drivers of Migration." *Journal of Ethnic and Migration Studies* 44 (6): 927–944.

Weidlich, Wolfgang, and Günter Haag. 2012. *Concepts and Models of a Quantitative Sociology: The Dynamics of Interacting Populations*. 14 vols. Berlin: Springer Science & Business Media.

(WPR) World Population Review. 2018. Accessed 7 April 2018. http://worldpopulationreview.com.

Exchange rate impact on output and inflation: A historical perspective from Zimbabwe

Nyasha Mahonye and Tatenda Zengeni

This research looks at the inflationary effect of currency devaluation and its contractionary effect on real output growth in Zimbabwe. The study uses quarterly data from 1990 to 2006 and utilizes the Johansen co-integration regression test and vector error correction method (VECM); and examines the short run and long run relationship between exchange rate, inflation and real output. The study finds that firstly, in both the short run and long run, fluctuations in the real exchange rate are significant on real output growth and expansionary in both periods. Secondly, the findings of the study suggest that exchange rate fluctuations are neither inflationary nor deflationary in Zimbabwe in the short run. Lastly, the result of the long run supports our hypothesis that devaluation of real exchange rate is inflationary. It implies that the weakening of domestic currency as part of the exchange rate liberalization policy is an incentive to Zimbabwean exporters and has potential economic growth gains though, in the long run, it has inflationary effects.

Introduction

The current research looks at the connection between exchange rate and domestic prices, and its potential contractionary[1] impact on real output in Zimbabwe between 1990 and 2006. Exchange rate movements play an anchor role in the country's export performance, which is expected to have an impact on output growth and general prices. Exchange rate measures competitiveness of goods and services produced abroad relative to those produced locally (Hassan 2014; Bahmani-Oskooee and Kandil 2007). The volume of exports normally increases when the income level abroad increases at higher level than relative prices (Hassan 2014; Bahmani-Oskooee and Kandil 2007). Exports are expected to increase when the foreign are higher than domestic prices. The increase in exports reflected by an increase in real gross domestic product following depreciation or devaluation is called the 'devaluation expansionary[?] hypothesis' whilst the decline in exports is termed the 'devaluation contractionary hypothesis'.

Many developed countries implemented the market-based exchange rate system in the 1970s and this resulted in uncertainty in exchange rates management (Bahmani-Oskooee and Hegerty 2007; Florian 2011). Research on appropriate exchange rate policy in low income countries is still ongoing among economist and is beyond the scope of this study; see the study by Bahmani-Oskooee and Hegerty (2007) and Florian (2011) for a discussion on the exchange rate regime and economic activities. Exchange rate policy is meant to tame the volatility of currency prices. Any movements in exchange rate might impact the real sector (Kandil and Mirzaie 2008; Ayubu 2013). This study seeks to provide evidence on lessons from the political economy case of Zimbabwe in relation to the potential impact of exchange rate.

The literature notes that the demand and supply channel determines these effects (Oskooee and Wang 2007). On the demand side, economic theory postulates

that devaluation[3] or depreciation could have a positive impact on production through stimulating the tradeable sector. On the supply side, it is argued that currency devaluation has the potential to raise the costs of imported raw materials and thus reduce production of real output. The arguments from the demand and supply side therefore attest that devaluation or depreciation has a potentially contractionary effect; otherwise, it could be expansionary.

This study looks at the link between exchange rate, output and inflation in Zimbabwe. Zimbabwe provides an ideal setting for studies on the connection between currency movements and real sector (see the discussion on real sector development in the next section). For example, in the past three decades, exchange rate volatility has persisted even after the dollarization policy in 2009 (Kararach and Otieno 2016). Previous studies on Zimbabwe have only investigated exchange rate pass through effect (see Mlambo and Dyrewall 1994; Ndhlela 2012). To the best of our knowledge, none of them has attempted to uncover the potential link between domestic currency movements and real sector in Zimbabwe. The period of 1990–2006 was chosen given that this period included an economic growth period (1990–1997) and an economic crisis period (1998–2006). The Zimbabwean economy experienced hyperinflation in the period of 2007 through to 2008 and inclusion of this period could distort the empirical analysis. The period of 2009 to the present (2018) saw implementation of a multicurrency system and hence the cessation of the Zimbabwean currency. This study seeks to provide evidence on lessons from the political economy case of Zimbabwe. The only credible study was done in 2016 by Kararach and Otieno (2016) but various studies in this book did not look at the potential connection between exchange rate and real sector. Accordingly, lessons from the political economy of Zimbabwe need to be interpreted within a historical perspective. Our research provides such a case study.

Our findings confirm that exchange rate devaluation or depreciation was expansionary in the case of the Zimbabwean economy between 1990 and 2006. Furthermore, our results show that currency weakening had an inflationary effect in the long run. The same results were found by Onafowora and Owoye (2008) for the case of Nigeria. The error correction results show the tendency of both real output and inflation to correct any short run disequilibrium that may arise. Our sensitive analysis tests were correlation tests (see Table A1 for details of these results), and ordinary least squares (OLS) for details of these results see Table A2. The same as the result by Kandil and Mirzaie (2008), our correlation tests attest to a positive relationship between currency depreciation and inflation whereas there is a negative connection between currency depreciation and real output. Ordinary least squares estimations attest to the positive relationship between exchange rate and inflation whereas a negative relationship was found between exchange rate devaluation and real output. On another note, we failed to control for potential structural factors such as the drought period of 1992, cyclone Eline of 2000 (floods), the partial drought period of 2005 and the potential effect of the land reform policy of 1999 on real output and food inflation.

Our results are robust, though, since we utilized the vector error correction model (VECM) that was used by the main empirical studies highlighted in Table 1. There are consistent findings between the Johansen co-integration test and error correction estimates, the correlation statistics and OLS estimations. Lastly, the variables included are mostly utilized by other studies that looked at this matter and we tested for stationarity and structural breaks as reported in the section on the empirical test findings. The rest of the paper is organized as follows. The next section looks at the overall trends in inflation, exchange rate and output during the period of 1990 through to 2006. The section thereafter presents the literature review. This is followed by the section that discussed the econometric model. The penultimate section presents the main empirical test findings while the final section provides concluding remarks and policy implication insights.

Trends on exchange rate, inflation and output performance in Zimbabwe

Over the past years, Zimbabwe has run a managed float exchange rate system of the Zimbabwean dollar to the US dollar. In our case, movements in real exchange rate can be deduced by observing the parallel market, which reveals that the Zimbabwean-US dollar value has been deviating from the market exchange rate in response to market forces.

During the Economic Structural Adjustment Programme (ESAP) period, the targeted inflation was 10% by 1995 (Zimstats Statistical Book 1995). This was to be addressed by monetary and credit policy reforms by liberating the credit and money market. Figure 1 shows the inflation and exchange rate trajectory in Zimbabwe between 1990 and 2006. Inflation was on an upward trend during the first years of reform; it rose from 12.4% in 1990 to 42.1% in 1992 before coming down to 27.6%

in 1993. The decreasing trend on inflation started in 1994 and continued until 1997, pointing to the impact of monetary reforms (Zimstats Statistical Book 2006).

The downward trend in price levels, however, reversed from 1998 onwards as the country started experiencing a hyperinflation period rising from 18.8% in 1997 to 133.2% in 2002. The annual official exchange rate and the price level for the period January 1990 to 1994 tended to trend together as shown in Figure 1. Thus, when the official exchange rate was weakening, inflation tended to increase. The exchange rate and inflation trended together in the period 1991–2000 (Figure 1). Inflation was increasing prior to August 1993, but immediately after a large devaluation it developed an upward shift in its trend.

Moreover, when the exchange rate declined at the end of 1992, inflation followed a similar trend (see Figure 1). During the middle of 1993, the rate of devaluation again exceeded inflation for several months and, as expected, inflation picked up in the beginning of 1994. In previous research (see Mlambo and Durewall 1994), devaluations that were noted to be larger than the rate of inflation eventually increased inflation. Furthermore, the impact of devaluations seems to have been greater and more rapid after the introduction of ESAP in November 1990. This was to be expected since there was a cut in the coverage of foreign exchange rationing and the opening up of the trade account that made tradable goods respond more quickly to imports costs.

According to Reserve Bank of Zimbabwe (RBZ) estimations, the net impact of the devaluation would result in the inflation rate increasing up to 23% at the beginning of year 1994 and would then decline to 15% by year end (RBZ, Sept 1995). As was predicted, inflation started rising and reached 23.5% in March 1994. Figure 1 shows that during the inception of ESAP, the inflation rate and the real exchange rate maintained stable movements with low fluctuations. The inflation rate was always trending above the official exchange rate during the whole period. On the same graph (Figure 1), the inflation rate took an upward trend in 2003 before taking a slump in the year 2004. During these two one-year periods, the real exchange rate maintained a steady trend, though it reached peak level in the second quarter of 2006, with inflation was slowing down in the same period. Both the official exchange rate and inflation adopted an upward trend thereafter.

Similarly, nominal GDP in Zimbabwe rose from $3.2 billion in 1980 to $14.7billion in 1990.The increase reflects increases in prices over this period, since real GDP rose from $3.2 billion in 1980 to $4.4 billion in 1990, an increase of about 3.75%. The fluctuations in real GDP are closely tied to rainfall patterns. For example, the three years with negative real growth rates (1983, 1987 and 1992) were the three driest seasons of the period. The close relationship between GDP and rainfall patterns is important in Zimbabwe given that the economy is agro-based. The goal under the ESAP was to raise gross investment from less than 20% of GDP in the late 1980s to around 25% of GDP by 1995.This was necessary to make it possible to achieve real GDP

Table 1: A summary of the systematic literature review on exchange rate, inflation and output: African and Asian countries' perspective.

Author, year	Country	Objective	Methodology used	Main outcome variables considered	Results/Findings	Data source used	Any comments
Alagidede and Ibrahim (2017)	Ghana	Determinant of exchange rate movements.	GARCH Model	Exchange rate, inflation, gross capital formation, real interest rate, money supply, government expenditure, terms of trade, and economic growth.	Excessive exchange rate has detrimental effect on real output.	Bank of Ghana and Datastream	The researcher did not factor the short run and long run effect of exchange rate movements.
Mbaluwa (2015)	Zimbabwe	Stock market performance, interest rate and exchange rate.	Vector Error Correction Model (VECM)	Exchange rate, interest rate (Treasury Bills) and stock market returns.	Pre-hyperinflationary period stock market performance influenced exchange rate movements and during hyperinflation exchange rate and interest rates had positive impact on stock market performance.	Reserve Bank of Zimbabwe (RBZ) statistics	This study neglected the link between exchange rate, inflation and real output.
Akinbobola (2012)	Nigeria	Co-movement between money supply, inflation and exchange rate.	Vector Error Correction Model (VECM)	Money supply, inflation, exchange rate, foreign prices and real output growth.	Money supply and exchange rate has long run inverse pressure inflationary whilst real output growth and foreign prices have direct effect on inflation.	International Financial Statistics (IFS)	This study is comprehensive enough but did not specifically look at the impact of exchange rate on prices and real output.
Masunda (2011)	Zimbabwe	Exchange rate misalignment and sectoral output performance.	Feasible Generalized Least Squares Panel Regression	Real exchange rate, mining output, manufacturing output and agricultural output.	Exchange rate undervaluation between 1980 and 2003 resulted in negative impact on sectoral performance in Zimbabwe.	Reserve Bank of Zimbabwe (RBZ), World Development Indicators (WDI), Central Office Statistics of Zimbabwe, IFS and Penn World tables 6.2	This study neglected its aggregate impact on output and its potential impact on price levels in Zimbabwe.
Basnet and Upadhyaya (2015)	ASEAN-5	Oil price shocks on output, prices and real exchange rate.	Structural VAR Model	Oil prices, real output, price levels and real exchange rate.	The macroeconomic variables of 5 Asian countries have common trends in the long run.	US Energy Information Agency Website, World Development Indicators and IFS	The research looked at the impact of oil prices on major macroeconomic variables but did not check the potential impact of exchange rate on output and inflation.
Rana and Dowling (1985)	Asia	Exchange rate and inflation.	Ordinary Least Squares (OLS)	Exchange rate, money supply and inflation.	Exchange rate movements and money supply growth does not affect inflation, but imported prices do influence domestic prices.	IMF's International Financial Statistics (IFS and Direction of Trade	This study looked at impact of exchange rate fluctuations on inflation only and neglected potential impact on real output.
Ojo and Alege (2014)	Sub-Saharan Africa (SSA)	Exchange rate and output.	System GMM	Exchange rate, output, government expenditure, FDI, interest rate and CPI.	There is a bidirectional and long run relationship between exchange rate and output for 40 SSA countries.	African Development Bank and IFS CD Room, 2008	Used a comprehensive econometric model but did not look at the potential relationship between exchange rate and inflation.
Bleaney and Greenaway (2001)	Sub-Saharan Africa (SSA)	Terms of trade, exchange rate on growth and investment.	Fixed effects panel regression model	Real GDP growth, gross capital formation, terms of trade, and exchange rate	Growth is negatively affected by terms of trade instability whilst investment is negatively affected by exchange rate instability.	IMF World Economic Outlook database	Neglected the dimension of potential impact of exchange rate on inflation.

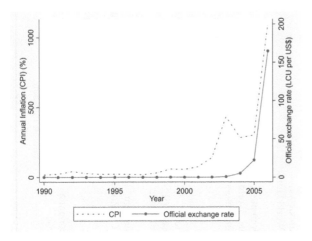

Figure 1: Trends in inflation and official exchange rate.
Source: World Bank Development Indicators, 2018

growth rates in the range of 5%, implying growth in real per capita GDP of close to 2% per year.

It is however, worth noting that according to provisional Zimstats estimates (1992), real GDP grew by a healthy 4.9% in 1991. This was the first year of the reform programme and, in spite of the fact that rainfall was below average, real GDP per capita grew by about 1.8%. This might imply that drought was not the sole contributor to fluctuations in real GDP and prices; hence, the need to consider the potential impact of domestic currency movements.

Zimbabwean dollar movements for the period 2002–2006[4]

The Zimbabwe dollar depreciated by an average 1.4% against major trading partner currencies in September 2002, as per publication by the Reserve Bank of Zimbabwe (RBZb 2003). Losses were recorded against the euro (4.7%), the Japanese yen, (4.5%), the South African rand, (0.3%) and the Botswana pula, (0.1%). During the same period, the local unit appreciated by 1% and 0.6% against the pound sterling and the franc, respectively (RBZb 2003). Given this trend, the auction system was introduced into the exchange rate determination in 2004 (Muñoz 2007). Since the inception of the auction system, the weighted average auction rate depreciated by 26.8% from Z$4196.61 per US$1 in Auction 1 (12 January 2004) to Z$5729.31/US$1 in Auction 96 (30 December 2004). On a monthly basis, the local unit depreciated by 1% from Z$5638.87 per US$ in November 2004 to Z$5696.031/US$ in December 2004. The Zimbabwe dollar depreciated against all major trading partner currencies by an average of 3.8% between November and December 2004. The observed movements came from the South African rand, 6.2%, Botswana pula, 8.3%, pound sterling, 5.3%, the euro, 4.1%, Swiss franc, 3.5% and the US dollar, 1%.

The auction system of exchange rate management continued to operate in 2005. In 2005, the local unit depreciated to Z$77,964.6 against the US$ by December 2005 from Z$64,282.61 recorded at the end of November 2005. During the same period, the Zimbabwe dollar depreciated by an average of 22.9% against all the other major

trading currencies. Again, the losses were recorded against the South African rand, 26.5%, Botswana pula, 24.2%; British pound, 22%; Canadian dollar, 23.2%, euro, 21.9% and the Japanese yen, 20.9%.

Related literature

Several researchers have examined the connection between exchange rate, domestic prices and output under distinct regimes of exchange rate. One of the most comprehensive multi-country studies by Ghosh et al. (1996) studied the effects of nominal anchor exchange rate regime on inflation and growth. The same approach was taken by Heravi (2015). Ghosh et al. (1996) found that the level and variability of inflation was noticeably lower under a fixed exchange rate than under a floating exchange rate regime. This implies that the positive association between exchange rate flexibility and inflation stressed in this study might not be monotonic. A study of transition economies by Domac, Peters, and Yuzefivich (2001) demonstrated that exchange rate regime has an effect for inflation performance. Furthermore, the findings of Agenor, McDermott, and Prasad (1997) show that exchange movements are a key variable of fluctuations in inflation in many developing countries. The result is consistent with the empirical findings of Kumar and Dhawan (1991) in their study of determinants of China's price dynamics.

Nonetheless, empirical findings underscore that the policy variables such as money supply, government spending as well as other variables such as net exports influence real national income and prices. These policy variables have different effects on growth under different exchange rate regimes. For example, it has been noted that monetary policy could help to minimize volatility in the exchange rate that may induce speculative attacks (Bahmani-Oskooee and Mirzaie 2000). This might result in fluctuation in output and prices. For instance, empirical studies from 'Algeria, Colombia, Cyprus, Ecuador, Guatemala, Honduras and India find evidence that non-neutral effects of anticipated monetary effects are evident by the positive effect on real output growth' (Kandil 2004, 96). In Kandil (2004), non-neutral effects of anticipated government spending shifts are evident by statistically significant positive impact on real output growth. In addition, Kandil (2004) asserts that anticipated demand shifts can raise the price level with corresponding real effective exchange rate falling and this has an expansionary effect on output.

Kandil (2004) found that anticipated monetary policy changes induce inflationary effect. This was evidenced by the significant response of price to monetary changes in the study of Honduras, Iran, Jordan, Malaysia, Cyprus, Ghana, and Sri Lanka (32% of the sample) Kandil (2004). The inflationary effect of projected government spending pattern was also evident in Algeria, Costa Rica, Greece, Honduras, Korea, Peru, and Turkey (32% of sample). The study by Rogers and Jenkins (1995) and Engel (1999) found an important role for the non-traded goods price in accounting for Canada–USA real exchange rate movements compared to the European countries–USA cases. Empirical studies on emerging markets shows that the countries which have allowed for a floating exchange rate encountering greater price fluctuations than countries that followed fixed exchange rates. Calvo (2000) applied similar logic to argue that a floating exchange rate can have destabilizing effects on emerging markets. The study found that shocks in exchange rate policy, which are captured by exchange rate fluctuations, in turn, cause other variables like prices, and possibly real GDP, to change.

Kandil and Mirzaie (2003) argued that many developing countries have a high variability of exchange rate in which its anticipated value was observed to generate adverse effects in the form of higher price inflation and large output contraction. In support of these findings, some studies have shown that although boosting net export through currency devaluation is desirable, it is crucial to ensure a concurrent increase in productive capacity to cope with the increased demand without accelerating price inflation (Kandil 2004). It is equally important to increase production capacity in import industries to reduce adverse effects of currency fluctuations on price inflation and the output supplied.

Most of previous studies on cross-country analyses used pooled time series, producing mixed results. For instance, while Nunnenkamp and Schweickert (1990) rejected the hypothesis of contractionary devaluation, Sheehey (1986) found that devaluation has a negative impact on output. Edwards (1986), on the other hand, concluded that devaluations have both a negative effect and a neutral effect on output in the short run and the long run respectively. The results from the pooled time series analyses studies are not robust since they consist of a very heterogeneous set of countries in which there are different growth effects of devaluation among the countries studied. Moreover, since the properties of the time series differ across the countries, in this study we differ from these previous works with respect to the identified properties of data because it focuses on one country analysis. The current study seeks to provide evidence on lessons from the political economy case of Zimbabwe; the only credible study was done in 2016 by Kararach and Otieno (2016); various studies in this book did not look at the potential connection between exchange rate and real sector. Accordingly, lessons from political economy of Zimbabwe have to be interpreted in historical perspective and our research provides one of such case study. For more previous empirical evidence on this matter, see Table 1 that has a summary of literature from African countries and Asian counterparts.

Empirical strategy

Lessons from the theoretical and empirical literature led this study to adopt Kandil and Mirzaie's (2002) model, stated below:

Long run model for output

The reduced form equation is expressed in logs to capture data variability,

$$RGDP_t = \alpha_0 + \alpha_1 M_t + \alpha_2 G_t + \alpha_3 RE_t + \alpha_4 NX_t$$
$$+ \varepsilon_t \quad (1)$$

The dependent variable is Real output, GDP_t which theoretically can be determined by the money supply,

M_t, which approximates monetary policy; by the government expenditure, denoted by G_t; by the real exchange rate denoted by RE_t; NX_t is net exports which approximates the activities in the current account. Unobserved changes in output are captured by ε_t.

These are expected signs for the output equation:

$$\alpha_1 > 0, \; \alpha_2 < 0, \; \alpha_3 < 0, \; \alpha_4 > 0$$

Theory posits that the coefficient of money supply to be positive; the coefficient of government spending is expected to be negative. Also, the expected sign of real exchange rate is negative (see Heravi (2015)) and, finally, the expected sign of the coefficient of net exports is positive Inferences of the expected signs are obtained from the literature review (see Kandil and Mirzaie (2002)). The model in equation (1) was tested for both long and short run relationships, following the suggestion by Engel and Granger (1987).

Short run model for output

$$\Delta GDP_t = \alpha_0 + \sum_{K=1}^{n1} \alpha_{1K}\Delta RGDP_{t-k} + \sum_{k=0}^{n2} \alpha_{2K}\Delta M_{t-k}$$
$$+ \sum_{K=0}^{n3} \alpha_{3K}\Delta G_{t-k} + \sum_{K=0}^{n4} \alpha_{4K}\Delta RE_{t-k} +$$
$$\sum_{K=0}^{n5} \alpha_{5K}\Delta NX_{t-k} + \beta \varepsilon_{t-1} + \omega_t \quad (2)$$

Model (4.1) adopted from Bahmani-Oskooee and Kandil (2007).

Real GDP depends on macroeconomic policy in the long run; the deviation between the right and left side of equation (1) as measured by ε_t, should reduce. The movement from short run to long run equilibrium is measured by the lagged value of ε_t as stipulated by the empirical model (4.1) (Bahmani-Oskooee and Kandil 2007). A negative and significant coefficient for ε_{t-1} indicates that variables are converging towards their equilibrium, that they co-integrated (Bahmani-Oskooee and Kandil 2007).

In the event that some variables are I(1) and some I(0), we follow the study by Pesaran, Shin, and Smith (2001), by estimating equation (2) as stipulated by specification in equation (3):

$$\Delta GDP_t = \alpha_0 + \sum_{K=1}^{n1} \alpha_{1K}\Delta RGDP_{t-k} + \sum_{K=0}^{n2} \alpha_{2K}\Delta M_{t-k}$$
$$+ \sum_{K=0}^{n3} \alpha_{3k}\Delta G_{t-k} + \sum_{K=0}^{n4} \alpha_{4K}\Delta RE_{t-k} + \beta_0 GDP_{t-1}$$
$$+ \beta_1 M_{t-1} + \beta_2 G_{t-1} + \beta_3 RE_{t-1} + \omega_t \quad (3)$$

Model (4.2) adopted from Bahmani-Oskooee and Kandil (2007).

Equation (3) is somewhat different from a standard vector autoregressive model (VAR) (Bahmani-Oskooee and Kandil (2007). This model includes a linear combination of lagged variables themselves, rather than the lagged of error term (Bahmani-Oskooee and Kandil 2007). Pesaran, Shin, and Smith (2001) proposed applying the familiar F-test for joint significance of the lagged level variables. According to Bahmani-Oskooee and Kandil (2007) 'If the lagged levels are jointly significant, there is a single level relationship between the dependent variables and the independent variable (that is they adjust jointly towards full employment)'.

The F-test that Pesaran, Shin, and Smith (2001) proposed has new critical values that they tabulate using the Monte Carlo experiment (Bahmani-Oskooee and Kandil 2007). By assuming all variables to be I(1) they provide an upper bound critical value and are I(0), a lower bound critical value is provided (Bahmani-Oskooee and Kandil 2007). For co-integration, the calculated F-statistic should be greater than the upper bound critical value (Bahmani-Oskooee and Kandil 2007). Once co-integration is established, the long run effects of independent variables on the dependent variable are inferred by the size and significant of β_1-β_3 that is normalized by β_0 (Bahmani-Oskooee and Kandil (2007). The short run impacts in this model are captured by the magnitude of α_{2K}-α_{4K} (Bahmani-Oskooee and Kandil 2007).

The model for the general prices

A handful of studies that have investigated the association between the real exchange rate and inflation in least developed countries (LDCs) include Holmes (2002), Bautista (1980), Rana and Dowling (1985), and Bahmani-Oskooee and Malixi (1992). These studies employed a single equation model with the real exchange rate as one of several variables that explain inflation (Holmes 2002). In the case of Rana and Dowling (1985), Bahmani-Oskooee and Malixi (1992) and Holmes (2002), the impact of the real exchange rate is modelled according to lag structures based on Almon or Koyck specifications.

In general terms, the methodological approach by all the above studies has been to model inflation as (Holmes 2002):

$$\pi_t = \beta_0 + \beta_1 M_t + \beta_2 G_t + \beta_3 RER_t + \beta_4 NX_t + \varepsilon_t \quad (4)$$

where:

π_t = the sustained increase in general price level.

NX_t = net exports as defined by exports net imports.

M_t = notes and coins in circulation plus demand deposits.

G_t = Government spending; recurrent and capital expenditure.

RER_t = Zimbabwean dollar exchange rate denoted as number of Zimbabwean dollars per US$ adjusted to inflation.

The following are expected signs for the price equation:

$$\beta_0 > 0, \beta_1 > 0, \beta_2 > 0, \beta_3 > 0 \, or < 0, \beta_4 > 0$$

The expected sign on the real exchange rate is ambiguous, since depreciation reduces the real value of assets denominated in local currency but increases the real value of assets denominated in foreign currencies (Holmes 2002). If the former effect dominates the latter, with a constant money supply this leads to an increase in inflation (Holmes 2002). Government spending is expected to be positive as was evident from previous studies based on developing countries. Money supply and net exports are expected to positively influence prices as discussed in the literature evidence.

Data sources and transformation

Data for the analysis were obtained from Reserve Bank of Zimbabwe's Economic and Research Policy Enhancement Division. The variables include Real Gross Domestic Product (RGDP), inflation, real exchange rate and money supply. The data on government spending and net exports (as in Kandil and Mirzaie 2008) were obtained from the Ministry of Finance in Zimbabwe. The data on nominal exchange rate from 1997 to 2006 were obtained from Old Mutual Asset Management. The data were obtained in quarterly time series covering the period 1990 first quarter to 2006 fourth quarter, translating into 68 observations.

Since the data on net exports and government expenditure were not available in quarterly form in Zimbabwe, the study utilized the technique proposed by Lisman and Sandee (1964) to splice data to quarterly data from annual totals. This method has good property that the quarterly data (supported by Kandil and Mirzaie 2007; Esen and Ozata 2010) it generates sum to exactly the annuals totals. Table 2 show the descriptive statistics from the data.

Empirical results

We discuss econometric results using time-series quarterly data from Zimbabwe in this section, following the approach of Mahonye and Mandishara (2014). We discuss unit roots tests and then cointegration empirical results follow, as conducted by Mahonye and Mandishara (2014). The cointegration test is tested following the Johansen (1988) and Johansen and Juselius (1990) procedure.

Diagnostic analysis and stationarity tests

The estimated model was tested for normality, serial correlation, autoregressive conditional heteroskedasticity and specification. Diagnostic test results are estimated in the model and are reported in Table 3. Diagnostics tests show no issue presence in the data of autocorrelation or heteroskedasticity, and the model was found to be well specified.

Before an estimation of the model is done, it is important to consider the underlying properties of the process that generate the variables. This study undertook a test for the stationarity of variables to avoid estimating spurious relationships.

The current research however utilizes the Augmented Dickey Fuller (ADF) to test for the presence of unit roots in variables in our model. All tests were run at a 5% and 10% level of significance. This was adopted because their estimation procedures are easy to conduct and interpretation of its results can be easily followed. The trend and intercept were found to be significant; therefore, the study included it in the estimations. Thus, the ADF

Table 2: Descriptive statistics on the data.

	Inflation	Government expenditure	M1	Net exports	Real GDP	Exchange rate
Mean	461.6059	10.23386	5.920777	−22652.9	18.00206	5.6766195
Median	131.65	9.428556	4.694849	440.5470	18.01742	4.159985
Maximum	3450.1	15.7578	14.89137	8652	18.5035	15.84366
Minimum	50.5	7.399707	2.631889	−456770	17.789	1.948763
Std. dev	789.0688	2.537403	3.10491	82563.59	.103442	3.553707
Skewness	2.5948	.917555	1.215492	−4.253382	−430989	1.118812
Kurtosis	9.197915	2.658086	3.528223	20.07778	2.397858	3.224926

Table 3: Diagnostics test results.

Test	F-statistic	Probability
Breush-Godfrey test	162.4571	0.6641
Ramsey reset	1.448897	0.5052708
Normality test (Jarque-Bera)	1.965415	0.374296
Arch test	98.88331	0.2876

Durban Watson statistic=2.185380 R-Adjusted=0.835674

Table 4: Unit roots test statistics of variables in levels.

Variable	ADF(Lags)	Remarks
Inflation	−2.737614(2)	Non-stationary
Real GDP	−0.938717(2)	Non-stationary
Real exchange rate	−1.965032(2)	Non-stationary
Net exports	−3.09273(2)	Non-stationary
Government expenditure	−1.537780(2)	Non-stationary
Money supply (M1)	2.850461(2)	Non-stationary
Critical values 5%	−3.4790	
Critical values 1%	−4.1035	

Notes. The critical values were obtained from the E-views Econometric package and are equivalent to the critical t-values in the Augmented Dickey-Fuller distribution table in Enders (1995).

results are generally reported as those from the tests that were run on lag two.

Based on the results in Table 4, we reject the null hypothesis of unit roots for all the variables in levels using the ADF test at both the 1% and 5% level. Thus, we proceed and estimate unit roots based on data in first difference to induce stationary. The unit roots results of differenced variables are presented in Table 5.

As shown above, ADF tests can be used to test for the order of integration. In order to determine the number of unit roots in the series, we proceeded to difference the series once to induce stationarity. Our test therefore failed to reject the null hypotheses that there is presence of unit root in all the variables in the model. The study proceed and tested the presence of unit root for differenced variables and found that all variables are stationary and integrated of order 1.

Cointegration results (long-run relationship)

First, we test for optimal lag length that should be utilized. Given the limited number of observations in our study using quarterly data of 1990–2006, we imposed a maximum of 7 lags. This was selected based on each differenced variable and relying first on the Akaike Information Criterion (AIC). Second, we test for cointegration, without invoking the requirement that all variables in the models are non-stationary. The results are reported in Table 4.

Table 6 shows the outcome of both trace and the maximum eigenvalue tests. Based on both the trace and the maximum eigenvalue statistics, there are four cointegrating equations (CEs). Therefore, the rank is equal to 4. The four CEs are obtained based on a model with a linear trend in the data, an intercept and no trend in it.

Table 5: Unit roots test statistics of variables in first differences.

Variables	ADF (lags)	Remarks
Inflation	−5.876562(2)	Stationary
Real GDP	−2.714966(2)	Stationary*
Exchange rate	−7.830044(2)	Stationary
Net exports	−6.277310(2)	Stationary**
Government expenditure	−6.644421(2)	Stationary
M1	4.639362	Stationary
Critical values 1%	−2.5989	*
**Critical values 5%	−1.9455	*

Notes. The ADF test statistics reported here were derived from ADF tests that were run on one lag.
*Variables are stationary at 1%
**variables are stationary at 5%

Table 6: Johansen co integration rank tests.

Series: DLRGDP DINFL DLGE DLM1 DLRER DNX				
Lags interval: 1–7				
Eigenvalue	Likelihood ratio	5% critical value	1% critical value	Hypothesized No. of CE(s)
0.923864	369.0358	94.15	103.18	None **
0.791219	217.0967	68.52	76.07	At most 1 **
0.748404	124.6750	47.21	54.46	At most 2 **
0.379432	43.25910	29.68	35.65	At most 3 **
0.218195	15.10906	15.41	20.04	At most 4
0.009886	0.586181	3.76	6.65	At most 5

Notes. ** denotes rejection of the hypothesis at 5% (1%) significance level
L.R test indicates 4 cointegrating equation(s) at 5% significance level
DLRGDP – Differenced Real Gross Domestic Product, **DINFL** – Differenced Inflation, **DLGE** – Differenced Government Expenditure, **DLM1** – Differenced Money Supply (Notes and Coins in circulation), **DLRER** – Differenced Real Effective Exchange Rate and **DNX** – Differenced Net Exports

The likelihood ratio is greater than the critical value in the first four rows. This shows that the hypothesis of no cointegration is rejected, suggesting that there exists a long run relationship among the variables in question, thus also suggesting that we can formulate an error correction model (ECM), which would reveal the short run dynamics of the model. Cointegration implies that the integrated series never drift far apart from each other, meaning they maintain equilibrium.

Table 7 show results of the cointegration rank tests, and Eviews (statistical software used for empirical analysis) provides the estimates of the cointegration vector or relations. In the study, we are interested in the first r-estimates, where r is determined by the LR tests.

Since the aim is to estimate the inflation and Real GDP equation, long run restrictions are imposed such that cointegrating equation 1 (`CEs I) is normalized with respect to Real GDP as shown by the results in Table 7 and with respect to inflation as shown by Table 8.

The estimated long run Real GDP and inflation equation(s) are therefore specified as follows, by equation (5) and (6) respectively:

$$\text{Real GDP} = -44.9 + 1.38\,\text{real effective exchange rate}$$
$$+ 0.21\,\text{government expenditure}$$
$$+ 2.39\,\text{money supply} + 0.97\,\text{inflation} \quad (5)$$

$$\text{Inflation} = 13.7 + 1.47\,\text{real effective exchange rate}$$
$$- 1.82\,\text{government expenditure}$$
$$+ 0.34\,\text{Real GDP} + 0.22\,\text{money supply}$$
$$- 0.0021\,\text{net exports} \quad (6)$$

In the long run. government spending and monetary growths are significant in determining the output growth and price fluctuations. Government expenditure was found to positively impact on output growth. In the long run. real exchange rate has a positive effect on real output growth, and it has a positive and significant impact on general prices. The results of other variables are as per our expectations: real output growth was found to influence inflation positively and its coefficient was significant. This implies that an increase in output growth will be met with an increase in general prices, and an increase in general prices will result in an increase in output growth.

The error correction model (ECM): Short-run relationship

The positive and negative coefficients imply pushing the system away from equilibrium and back to equilibrium, respectively. If there is an economic shock that pushes inflation away from the equilibrium in Table 8, government spending and real output growth would adjust by pushing away the system from the desired equilibrium. For the same equation, monetary growth, net exports and real exchange rate adjustment coefficient, although negative and positive, are not significant. This means that if there is a shock that pushes inflation away from the equilibrium in Table 9, net exports, real exchange rate and monetary growth cannot correct the discrepancy. Thus, the devaluation of a Zimbabwean dollar has a neutral effect on inflation in the short run.

The negative coefficient on government spending for the output equation implies that the system is pushing back to equilibrium. If there is an economic shock that pushes real output growth away from the equilibrium in Table 9, government spending would adjust by pushing the system back into the desired equilibrium. For the same equation, inflation, real exchange rate and net exports have a positive coefficient, and this implies that if there is an economic shock on real output growth, the system will push away from the equilibrium. However,

Table 7: Normalized cointegrating coefficients: 1 cointegrating equation(s)-DLRGDP.

DLRGDP	DLGE	DLM1	DLRER	DINFL	C
1.000000	−0.21046	−2.3890	−1.38057	−0.97283	4.485945
	(24.6110)	(23.1767)	(7.92553)	(160.387)	
Log =likelihood	−0.305226				

Notes: **DLRGDP** – Differenced Real Gross domestic Product, **DINFL** – Differenced Inflation, **DLGE** – Differenced Government Expenditure, **DLM1** – Differenced Money Supply (Notes and Coins in circulation), and **DLRER** – Differenced Real Effective Exchange Rate

Table 8: Normalized cointegrating coefficients: 1 cointegrating equation(s)-DINFL.

DINFL	DLGE	DLM1	DLRER	DLRGDP	DNX	C
1.000000	1.82107	−0.2189	−1.46669	−0.3420	0.002115	−1.36569
	(5.62846)	(9.21491)	(1.73876)	(33.4721)	(0.00016)	

Log = likelihood   −532.3475

Notes: **DLRGDP** – Differenced Real Gross Domestic Product, **DINFL** – Differenced Inflation, **DLGE** – Differenced Government Expenditure, **DLM1** – Differenced Money Supply (Notes and Coins in circulation), **DLRER** – Differenced Real Effective Exchange Rate and **DNX** – Differenced Net Exports

Table 9: Short run error correction results (Inflation Equation).

Error correction:	D(DINFL)	D(DLGE)	D(DLM1)	D(DLRER)	D(DLRGDP)	D(DNX)
CointEq1	−6.625989	0.093963	−0.005988	−0.003643	0.000996	−719.9010
	(2.24143)	(0.03960)	(0.00626)	(0.10998)	(0.00026)	(1304.12)
	(−2.95614)	(2.37287)	(−0.95706)	(−0.03313)	(3.78260)	(−0.55202)

Notes: **DLRGDP** – Differenced Real Gross Domestic Product, **DINFL** – Differenced Inflation, **DLGE** – Differenced Government Expenditure, **DLM1** – Differenced Money Supply (Notes and Coins in circulation), **DLRER** – Differenced Real Effective Exchange Rate and **DNX** – Differenced Net Exports

Table 10: Short run error correction results (Output Equation).

Error correction:	D(DLRGDP)	D(DINFL)	D(DLGE)	D(DLM1)	D(DLRER)	D(DNX)
CointEq1	−0.314291	1971.726	−26.54369	0.751761	89.26641	1703956.
	(0.08418)	(716.263)	(12.6540)	(1.99936)	(35.1453)	(416741.)
	(−3.73345)	(2.75280)	(−2.09765)	(0.37600)	(2.53993)	(4.08877)

Notes: **DLRGDP** – Differenced Real Gross Domestic Product, **DINFL** – Differenced Inflation, **DLGE** – Differenced Government Expenditure, **DLM1** – Differenced Money Supply (Notes and Coins in circulation), **DLRER** – Differenced Real Effective Exchange Rate and **DNX** – Differenced Net Exports

though these variables are significant, they are not capable of bringing back real output growth to its long run equilibrium (Table 10).

The vector error correction method (VECM) has a special appeal in the model that includes real exchange rate. In countries where the purchasing power parity (PPP) holds, the real rate is stationary or I(0) where other variables are I(1). In the case of Zimbabwe, we found that using the parallel exchange rate, the PPP does not holds, since the real exchange rate is I(1).

In short run, movement in the real exchange rate is significant on real output growth and it was found to be expansionary. Over the long run, depreciation or devaluation of the exchange rate also had a positive effect on the real output growth, as evident from the positive and significant coefficient. In both the short run and the long run, the results were consistent on the effect of real exchange rate on real output. The results were so because as a producer anticipates sustained real depreciation and an increase in production, this provides a greater incentive to produce and promote exports, resulting in a long-lasting increase in output growth. This result however rejects our hypothesis that devaluation of the exchange rate is contractionary.

Monetary growth encourages real output growth in the long run; however, it was found to be insignificant in the short run (Bahmani-Oskooee and Kandil 2007). Government expenditure has a key impact on real GDP. Government spending was found to discourage real output growth since it is significant and negative. The results point to the weakness of the monetary and fiscal policy in generating a positive real GDP growth trajectory in Zimbabwe.

In the long run, the movements in general price level were found to encourage output growth. This is not consistent with the theory since a high level of inflation is a disincentive to producers of commodities. In our findings, this was evident by a significant long run relationship between real output and inflation. This does not attest to the economic decline experienced by the Zimbabwean economy during the 1997–2006 period. The decline in the real output was accompanied by persistent inflation during the period under investigation. However, the short run dynamics between real output growth and inflation was found to be positive and significant. This is same result found by Kandil (2006). As a result, inflation produced real output gains in the short run in Zimbabwe. This means that there is a certain level of inflation that produces real output growth in both the long run and short run.

The real exchange rate was found to be insignificant for short run determination of the general price levels. Our findings support the results obtained by Ghosh et al. (1996) that association between exchange rate and inflation variability is not monotonic but country specific. The findings in this study suggest that exchange rate movements were neither inflationary nor deflationary in Zimbabwe for the period 1997–2006 in the short run. The result shows that the devaluation had a neutral effect on inflation in the short run. The result of the long run supports our hypothesis that devaluation of the real exchange rate was inflationary. This was evidenced by a significant coefficient of the real exchange rate.

Monetary growth was found to be significant and positive in the long run. The result agrees with the Monetarist model that sees inflation as a monetary phenomenon. This result attests to the view that excessive monetary growth has an inflationary impact on the economy. However, the monetary growths were found to have no short-term dynamic effect on the general prices.

In the long run the real output growth was found to be significant and positive. This does not support our expectation that positive real output growth tends to stabilize general price levels. This is so because recurrent real output decline witnessed by the Zimbabwean economy between 1997–2006 was met with a phenomenal general price surge. However, in the short-term, real output growth was found to be inflationary. The expansionary effect of real exchange rate attested by the estimated results entails that the devaluation or depreciation will result in aggregate supply of exports exceeding aggregate demand. This means that the devaluation of the exchange rate as part of the exchange rate liberalization policy was an incentive to the export producing sector and made Zimbabwean exports less expensive and hence more favourable on the international market. It also worth noting that most Zimbabwean exports are agricultural products and minerals, most of which are exported in their raw state. Consequently, they usually fetch less export earnings and hence the export producing sector continued to shrink. Zimbabwe is also heavily reliant on imported raw materials and with the devaluation of the Zimbabwean dollar the expansionary effect of this policy was not sustainable.

Conclusion and policy implication/s

This study investigated the inflationary effect of currency devaluation and its contractionary effect on real output growth in Zimbabwe. The study used data from 1990 to 2006 and utilized the Johansen co-integration regression test; and ascertained the long run and short run relationship between exchange rate, inflation and real output. The results from the study are threefold. Firstly, fluctuations in the real exchange rate are significant on real output growth and were found to be expansionary in both the short and long run. Are the results you summarise here referring to the study period (and should therefore use the past tense) or are you making general observations/presenting universal facts (which require the present tense)? Secondly, the findings of the study suggest that exchange rate fluctuations are neither inflationary nor deflationary in Zimbabwe in the short run. Lastly, the long run results support our hypothesis that devaluation of real exchange rate is inflationary.

The expansionary effect of real exchange rate attested by the estimated results entails that the devaluation or depreciation will result in aggregate supply of exports exceeding the aggregate demand, which means during the period under review, devaluation of the currency was output expansionary. This means that the devaluation of the exchange rate as part of the exchange rate liberalization policy was an incentive to Zimbabwean exporters. It is also worth noting that most of Zimbabwean exports are commodities comprising agricultural products and minerals whose prices are determined on the international market, and they usually fetch less export earnings during commodity prices bust and hence the export producing sector continues to shrink. Zimbabwe also heavily relies on imported raw materials and with devaluation of the Zimbabwean dollar the expansionary effect of this policy is not sustainable as it increases production cost.

This implies that the devaluation of the exchange rate as part of the exchange rate liberalization policy is an incentive to Zimbabwean exporters and has potential economic growth gains though, in the long run, it has inflationary effects.

Disclosure statement

No potential conflict of interest was reported by the authors. The views expressed in this study and any perceived errors and omissions are the authors' and not those of the affiliated organizations.

Notes

1. Contractionary effect means devaluation of currency has a negative effect on Real Gross Domestic Product.
2. Expansionary effect means devaluation of currency has positive effect on Real Gross Domestic Product (Vaithilingam, Guru, and Shanmugam 2003).
3. Devaluation is the weakening of domestic currency against another country's currency. In this study we use it interchangeably with depreciation of exchange rate.
4. Data source for this section is Zimbabwe Statistics (ZIMSTATS) (1992; 1995, 2006) Statistical Yearbook and the Reserve Bank of Zimbabwe (2006).

References

Agenor, P. R., C. J. McDermott, and E. S. Prasad. 1997. "Macroeconomic Fluctuations in Developing Countries: Some Stylized Facts." *IMF Working Paper*, WP/99/35.

Akinbobola, T. 2012. "The Dynamics of Money Supply, Exchange Rate and Inflation in Nigeria." *Journal of Applied Finance and Banking* 2 (4): 117–141.

Alagidede, P., and M. Ibrahim. 2017. "On the Causes and Effects of Exchange Rate Volatility on Economic Growth: Evidence from Ghana." *Journal of African Business* 18 (2): 169–193.

Ayubu, V. S. 2013. "Monetary Policy and Inflation Dynamics: An empirical case study of Tanzanian Economy." Umea Universitet.

Bahmani-Oskooee, M., and S. W. Hegerty. 2007. "Exchange Rate Volatility and Trade Flows: A Review Article." *Journal of Economic Studies* 34 (3): 211–255.

Bahmani-Oskooee, M., and M. M. E. Kandil. 2007. "Exchange Rate Fluctuations and Output in Oil-Producing Countries: The Case of Iran" (No 7–113). *International Monetary Fund*.

Bahmani-Oskooee, M., and M. Malixi. 1992. "Inflationary Effects of Changes in Effective Exchange Rate: LDCs Experience." *Applied Economics* 24: 465–471.

Bahmani-Oskooee, M., and A. Mirzaie. 2000. "The Long-Run Effects of Depreciation of the Dollar on Sectoral Output." *International Economic Journal* 14: 51–61.

Basnet, H. C., and K. P. Upadhyaya. 2015. "Impact of Oil Price Shocks on Output, Inflation and the Real Exchange Rate: Evidence from Selected ASEAN Countries." *Applied Economics* 47 (29): 3078–3091.

Bautista, R. 1980. "Exchange Rate Adjustment under Generalized Floating Comparative Analysis among Developing Countries." *World Bank Staff Working Paper No436, October, Washington DC: World Bank*.

Bleaney, M., and D. Greenaway. 2001. "The Impact of Terms of Trade and Real Exchange Rate Volatility on Investment and Growth in Sub-Saharan Africa." *Journal of Development Economics* 65: 491–500.

Calvo, G. 2000. "Relative Rate Volatility and Economic Performance in Peru: A Firm Level Analysis." *Emerging Markets Review* 4 (4): 472–496.

Domac, I., K. Peters, and Y. Yuzefivich. 2001. "Does the Exchange Rate Regime affect Macroeconomic Performance? Evidence from Transition Economies." Washington DC: World Bank.

Edwards, S. 1986. "Are Devaluations Contractionary?" *The Review of Economics and Statistics* 68: 501–508.

Enders, W. 1995. *Applied Econometrics Time Series*. New York: John Wiley and Sons, Inc.

Engel, C. 1999. "Accounting for US Real Exchange Rate Changes." *Journal of Political Economy* 107: 507–538.

Engel, R. F., and C. W. J. Granger. 1987. "Co-Integration and Error Correction: Representation, Estimation, and Testing." *Econometrica* 55: 251–276.

Esen, E., and E. Ozata. 2010. "The Impacts of Tourism on Economic Growth: Analyzing the Validity of the Tourism Led Growth Hypothesis for Turkey with the ARDL Model." *Anadolu University Journal of Social Sciences* 17 (1): 43–51.

Florian, V. 2011. "Bilateral Exports from Euro Zone Countries to the US: Does Exchange Rate Variability Play a Role?" Discussion papers, Center for European Governance and Economic Development Research, No. 121, CeGE, Gottingen.

Ghosh, R. A., A. M. Gulde, J. D. Ostry, and H. Wolf. 1996. "Does the Exchange Rate Regime Matter for Inflation and Growth?" *Economic Issues 2*. Washington DC: International Monetary Fund.

Hassan, S. S. U. 2014. "Exchange Rate Volatility, UK Imports and The Recent Financial Crisis: Evidence from Symmetric ARDL and Asymmetric ARDL Methods." Doctoral Dissertation, University of Southampton.

Heravi, M. M. L. 2015. "Real Exchange Rate in Commodity Exporting Countries." Adam Smith Business School, University of Glasgow.

Holmes, J. M. 2002. "The Inflationary Effects of Effective Exchange Rate Depreciation in Selected African Countries." *Journal of African Economics* 11 (2): 201–218.

Johansen, S. 1988. "Statistical Analysis of Cointegration Vectors." *Journal of Economic Dynamics and Control* 12 (2–3): 231–254.

Johansen, S., and K. Juselius. 1990. "Maximum Likelihood Estimation and Inference on Cointegration – with Applications to the Demand for Money." *Oxford Bulletin of Economics and Statistics* 52 (2): 169–210.

Kandil, M. 2004. "Exchange Rate Fluctuations and Economic Activities in Developing Countries: Theory and Evidence." *Journal of Economic Development* 29 (1): 85–108.

Kandil, M. 2006. "On the Transmission Mechanism of Policy Shocks in Developing Countries." *Oxford Development Studies* 34 (2): 117–149.

Kandil, M., and I. A. Mirzaie. 2002. "Comparative Analysis of Exchange Rate Fluctuations on Output and Price: Evidence from Middle Eastern Countries." IMF Institute, Washington DC, USA.

Kandil, M., and I. A. Mirzaie. 2003. "The Effects of Exchange Rate Fluctuations on Output and Prices: Evidence from Developing Countries." John Carroll University, USA.

Kandil, M. and I. A. Mirzaie. 2007. "Consumption and Macroeconomic Policies: Evidence of Asymmetry in Developing Countries." *International Journal of Development Studies Issues* 6 (2): 83–105.

Kandil, M., and I. A. Mirzaie. 2008. "Comparative Analysis of Exchange Rate Appreciation and Aggregate Economic Activity: Theory and Evidence from Middle Eastern Countries." *Bulletin of Economic Research* 60 (1): 45–96.

Kararach, G., and R. O. Otieno. 2016. *Economic Management in a Hyperinflationary Environment: The Political Economy of Zimbabwe, 1980–2008*. Oxford: Oxford University Press.

Kumar, R., and R. Dhawan. 1991. "Exchange Rate Volatility and Pakistan's Exports to the Developed World, 1974–85." *World Development* 19: 1225–1240.

Lisman, J. H. C., and J. Sandee. 1964. "Derivation of Quarterly Figures from Annual Data." *Applied Statistics* 13: 87–90.

Mahonye, N., and L. Mandishara. 2014. "Stock Market Returns and Hyperinflation in Zimbabwe." *Investment Management and Financial Innovations* 11 (4–1): 223–232.

Masunda, S. 2011. "Real Exchange Rate Misalignment and Sectoral Output in Zimbabwe." *International Journal of Economics and Research* 2 (4): 59–74.

Mbaluwa, S. 2015. "Stock Market Performance, Interest Rate and Exchange Rate Interactions in Zimbabwe: A Cointegration Approach." *International Journal of Economics, Finance and Management* 4 (2): 77–88.

Mlambo, K., and D. Durewall. 1994. *Inflation and Stabilization Policies in Zimbabwe*. Harare: University of Zimbabwe Publications.

Muñoz, M. S. 2007. "Central Bank Quasi-Fiscal Losses and High Inflation in Zimbabwe: A Note" (No 7–98). *International Monetary Fund*.

Ndhlela, T. 2012. "Implications of Real Exchange Rate Misalignment in Developing Countries: Theory, Empirical Evidence and Application to Growth Performance in Zimbabwe." *South African Journal of Economics* 80 (3): 319–344.

Nunnenkamp, P., and R. Schweickert. 1990. "Adjustment Policies and Economic Growth in Developing Countries-Is Devaluation Contractionary?" *Weltwirtschaftliches Archiv* 126: 474–493.

Ojo, A. T., and P. O. Alege. 2014. "Exchange Rate Fluctuations and Macroeconomics Performance in Sub-Saharan Africa: A Dynamic Cointegration Analysis." *Asian Economic and Financial Review* 4 (11): 1573–1591.

Onafowora, O. A., and O. Owoye. 2008. "Exchange Rate Volatility and Export Growth in Nigeria." *Applied Economics* 40 (12): 1547–1556.

Oskooee, B. M. and Wang, Y. 2007. "The Impact of Exchange Rate Volatility on Commodity Trade between the U.S. and China." *Economic Issues* 12 (1): 31–51.

Pesaran, M., Y. Shin, and R. Smith. 2001. "Bounds Testing Approaches to the Analysis of Level Relationships." *Journal of Applied Econometrics* 16: 289–326.

Rana, P. B., and J. M. Dowling. 1985. "Inflationary Effects of Small But Continuous Changes in Effective Exchange Rates: Nine Asian LDCs." *The Review of Economics and Statistics* 67 (3): 496–500.

Reserve Bank of Zimbabwe. 1995, 2003, 2006. http://www.rbz.co.zw/statistics.

Rogers, J. H., and M. Jenkins. 1995. "Haircuts or Hysteries Sources of Movements of Real Exchange Rates." *Journal of International Economics* 38: 339–360.

Sheehey, E. J. 1986. "Unanticipated Inflation, Devaluation, and Output in Latin America." *World Development* 14: 665–671.

Vaithilingam, S., B. K. Guru, and Bala Shanmugam. 2003. "Bank Lending and Economic Growth in Malaysia." *Journal of Asian Pacific* 5 (1): 51–69.

(ZIMSTATS) Zimbabwe Statistics. 1992, 1995, 2006. Statistical Yearbook.

Appendix

Table A1: Correlation statistics.

	Exchange rate	Inflation	Real GDP	Money supply (M1)	Net exports	Government spending
Exchange rate	1.00					
Inflation	0.86*	1.00				
Real GDP	−0.32*	−0.50*	1.00			
Money supply (M1)	0.998*	0.86*	−0.33*	1.00		
Net exports	−0.35*	−0.34*	0.23*	−0.35*	1.00	
Government spending	0.94*	0.77*	−0.27*	0.94*	−0.32*	1.00

Notes: * significance at 10% level

Table A2: Ordinary least squares (OLS) regression model.

	Inflation		Real GDP	
	Coef	SE	Coef	SE
Exchange rate	0.62***	(0.09)	−0.05*	(0.02)
M1	−0.25*	(0.11)	0.02	(0.03)
Net exports	−0.00***	(0.00)	0.00	(0.00)
Government expenditure	−0.10	(0.05)	0.01	(0.01)
_constant	4.05***	(0.35)	18.05***	(0.08)
N	68		68	

Notes: ***Significant at 1% level; **significant at 5% level; *significant at 10% level. Coef = coefficient; SE = Standard error.

A mathematical formulation of the joint economic and emission dispatch problem of a renewable energy-assisted prosumer microgrid

Uyikumhe Damisa, Nnamdi I. Nwulu and Yanxia Sun

Operational planning of prosumer microgrids with solar and wind energy sources is quite a complex task considering the intermittency of these sources and energy import/export from prosumers. Reserve capacities which can be reliably provided by dispatchable sources like conventional generators (CGs) may be needed to ensure reliability of the grid. However, these sources produce emissions which have adverse effects on the environment. Hence, emission curtailment should be incorporated in the operational planning of microgrids with these generators. In this paper, a mathematical formulation for the joint economic and emission dispatch of a renewable energy-assisted prosumer microgrid is presented and solved using the CPLEX Solver in Advanced Interactive Multidimensional Modelling System (AIMMS). A modified microgrid test system is used as a case study in this work. Results show that incorporating an emission function in the objective of the operational dispatch formulation not only reduces emissions, but could be of advantage to customers as larger capacities of their behind-the-meter resources get the chance to provide grid ancillary services; however, it also puts a restriction on the profit that could be made from selling energy to the main grid during periods when energy prices are high.

Nomenclature

Sets

t, T	index for, and number of timeslots
g, G	index for, and number of CGs
p, P	index for, and number of prosumers

Parameters

Δt	Duration of a timeslot [hrs]
E_{Solar}^t	Aggregate solar output during timeslot t [kWh]
E_{Wind}^t	Wind farm output during timeslot t [kWh]
E_R^t	Required reserve capacity for the duration of timeslot t [kWh]
$E_{ProGen, p}^t$	Prosumer p's onsite generator output during timeslot t [kWh]
$E_{NDL,p}^t$	Energy required for prosumer p's non-deferrable load during timeslot t [kWh]
$E_{BatTxf,p}^{max}$	Maximum quantity of charge/discharge of prosumer p's battery per timeslot [kWh]
$E_{Batt,p}^{min}$	Minimum energy level of prosumer p's battery [kWh]
$E_{Batt,p}^{max}$	Maximum energy level of prosumer p's battery [kWh]
$E_{DL,p}$	Amount of deferrable load to be served in prosumer p per timeslot [kWh]
$E_{DL,p}^{total}$	Total amount of deferrable load in prosumer p [kWh]
$E_{OtherLoad}^t$	Total demand (excluding prosumer demand) in timeslot t [kWh]
$P_{c.gen,g}^{min}$	Lower limit of CG g's output [kW]
$P_{c.gen,g}^{max}$	Upper limit of CG g's output [kW]
r	Up/down ramping coefficient
a_g, b_g & c_g	Coefficients of CG g's generation cost [\$h^{-1}]/ [\$(kWh)$^{-1}$]/[\$(kW^{2}h)$^{-1}$]
$C_{c.gen,g}^t$	Cost of generation of CG g for the duration of timeslot t [\$]
C_{sell}^t	Selling cost of energy [\$(kWh)$^{-1}$]
C_{buy}^t	Buying cost of energy [\$(kWh)$^{-1}$]
n_p	(Dis)Charge efficiency of battery in prosumer p

Variables

$E_{c.gen,g}^t$	CG g's output for the duration of timeslot t [kWh]
E_{buy}^t	Energy scheduled to be bought from main grid for the duration of timeslot t [kWh]
S_{Buy}^t	Energy bought status [Yes (1), No (0)]
E_{Sell}^t	Energy scheduled to be sold to the main grid for the duration of timeslot t [kWh]
S_{sell}^t	Energy sold status [Yes (1), No (0)]
$S_{c.gen,g}^t$	ON/OFF state of CG g in timeslot t
$S_{DL,p}^t$	ON/OFF state of deferrable load in prosumer p in timeslot t
$E_{GridPro,p}^t$	Energy trade between the main grid and prosumer p during timeslot t [kWh]
$E_{BatTxf,p}^t$	Amount of charge/discharge of prosumer p's battery during timeslot t [kWh]
$P_{c.gen,g}^t$	CG g's power output during timeslot t [kW]
$E_{Bat,p}^t$	Energy content of prosumer p's battery during timeslot t [kWh]
$R_{c.gen,g}^t$	Reserve capacity from CG g during timeslot t [kWh]
$CR_{c.gen,g}^t$	Price of reserve capacity from CG g for the duration of timeslot t [\$]
$R_{Bat,p}^t$	Reserve capacity from prosumer p's battery in timeslot t [kWh]
$CR_{Bat,p}^t$	Price for reserve capacity from prosumer p's battery in timeslot t [kWh]
EM_g	Emissions produced per unit of power generated by CG g [kg]
EMR_g	Emissions produced per unit of reserve from CG g [kg]
$E_{MiMaTxfGrid}^{max}$	Maximum possible energy transfer between micro and main grid
$DevE_{Wind}^{-,t}$	Maximum downward deviation from wind power forecast in timeslot t [kWh]
$DevE_{Solar}^{-,t}$	Maximum downward deviation from solar power forecast in timeslot t [kWh]
$DevE_{Demand}^{+,t}$	Maximum upward deviation from demand forecast in timeslot t [kWh]

$DevE_{ProGen,p}^{-,t}$ Maximum downward deviation from prosumer
                       p's local generation forecast in timeslot t [kWh]
$DevE_{ProDem,p}^{+,t}$ Maximum upward deviation from prosumer p's
                       demand forecast in timeslot t [kWh]
NB. 1 kWh   3600000 J

Introduction

The major adverse effect of excessive release of green-house gases into the atmosphere, known as global warming, continues to attract the attention of the public. In the bid to reduce these harmful emissions, some directives like the Large Combustion Plant Directive, the Kyoto Protocol and the National Emissions Ceilings Directive have been put in place. To effectively curb these emissions, attention must be given to the electricity industry which is one of the major producers of these gases.

A low capital approach to emission reduction in the power industry is the use of an economic and emission dispatch model for the operational dispatch of a grid; it is also an approach that is relatively easy to implement (Nwulu and Xia 2015a).

Over the years, the economic dispatch problem has evolved; in addition to the dispatch of supply resources, flexible loads and storage facilities can now be simultaneously dispatched. In general, energy management or operational dispatch may be carried out by optimally scheduling supply resources CGs, demand (deferrable loads), storage (batteries) or any combination of these (Pal et al. 2016). In Ahn and Peng (2013) and Huang, Yao, and Wu (2014), energy management was carried out by scheduling the output of CGs. Deferrable load scheduling was employed in Chen, Shroff, and Sinha (2013), Farimani and Mashhadi (2015) and He et al. (2016) for effective energy management. In Bestehorn and Borsche (2014), power supply-demand balance was achieved by switching both producers and consumers. The authors, Su and El Gamal (2013) analyzed the use of energy storage in realizing power balance. In Nguyen and Crow (2016), the dispatch of generation and batteries was coordinated to minimize the expected operating cost of a microgrid, but contrary to the norm, power balance was not considered as a hard constraint; it was enforced with high probability. None of the abovementioned papers used the method of coordinated scheduling of supply, demand and storage components.

Farimani and Mashhadi (2013), Salinas et al. (2013), Zhang, Gatsis, and Giannakis (2013) and Pal et al. (2016) are closely related to this work as they adopted the approach of joint scheduling of supply, demand and storage. In Pal et al. (2016), a prosumer smart grid architecture consisting of a CG, multiple renewable energy generators (each having a storage unit attached), elastic and inelastic loads, and a connection to an external energy market was considered. The authors minimized a cost function consisting of conventional generation cost, battery degradation cost and net cost of power exchange with the external energy market, whilst ensuring power balance. The authors, Farimani and Mashhadi (2013) considered a hybrid system consisting of diesel, wind and solar generators, a storage unit, and dispatchable and undispatchable loads. They minimized an objective function consisting of diesel generation cost and cost of not supplying next-day load. In Salinas et al.

(2013), a distribution network of energy users each with a renewable energy source with battery storage and a connection to a power grid was studied. An optimization model that minimizes the cost of satisfying customers' total load demand was formulated.

None of the studies by Farimani and Mashhadi (2013), Salinas et al. (2013) or Pal et al. (2016) incorporated reserve scheduling into their operational dispatch formulation. In Zhang, Gatsis, and Giannakis (2013), however, provision was made for a spinning reserve capacity, but this was to be provided by CGs alone. The use of other grid components, aside from CGs, to deliver ancillary services, has been shown to reduce grid operating costs (Zakariazadeh, Jadid, and Siano 2014a). In Zakariazadeh, Jadid, and Siano (2014a), an energy and reserve dispatch formulation for a smart distribution grid was presented. The formulation was developed as a multi-objective optimization problem with two objectives, viz. cost and emissions.

Active consumers, known as prosumers, have begun to emerge in recent times. Prosumers are consumers of electricity who also have onsite generators like diesel generators, solar photovoltaic (PV) systems and wind turbines, and hence can feed electricity back to the grid or share it with other customers (Zafar et al. 2018). Their emergence is due to dwindling prices of renewable energy systems, favourable government policies, environmental pollution concerns, among other factors. With PV systems being one of the most preferred onsite generation technologies (Couture et al. 2014), many prosumers adopt this technology. The influx of prosumers with such unpredictable renewable energy sources complicates the task of grid operators, especially in cases where prosumers are permitted to unsolicitedly feed excess unused local generation to the grid. However, in addition to being key stakeholders of the future grid, prosumers can also play a crucial role in peak demand management (Zafar et al. 2018). An operational energy and reserve dispatch model for a prosumer microgrid was developed in Damisa, Nwulu, and Sun (2018). Two large prosumers, each with a renewable energy generator, a battery storage facility and flexible loads, were assumed to be connected to the microgrid. Their batteries and flexible loads were used by the microgrid operator (MGO) to achieve grid power balance and provide ancillary services. The operating cost of the microgrid was seen to reduce with the participation of these resources.

In this paper, the mathematical model in Damisa, Nwulu, and Sun (2018) is extended to include CG emission minimization. The resulting mathematical formulation is a multi-objective optimization problem with two objectives, namely, operating cost and emission function. It is solved using the weighing method of handling multi-objective optimization problems. Three weight combinations are investigated. Firstly, equal weights are assigned to both objective functions; the second combination effectively retains the operating cost alone, and the last combination retains the emission function alone. For each of these combinations, an optimal energy and reserve schedule is generated. Furthermore, a comparison of the results for each weight combination is reported.

The rest of the paper follows the following structure: A brief description and schematic of the microgrid

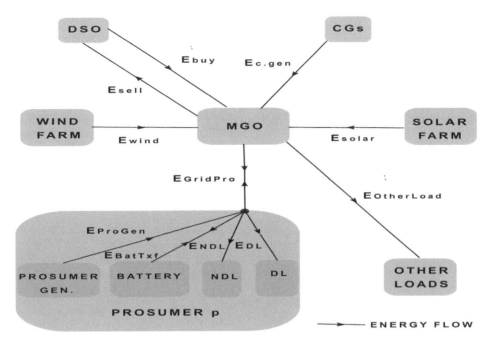

Figure 1: Diagrammatic representation of microgrid configuration.

configuration is given in the next section. The mathematical model of the grid's dynamics is presented in the section thereafter. Then follows the section containing a brief description of the simulation set-up, with the results presented next. A discussion of the results, together with policy recommendations is presented in the penultimate section, with the final section concluding the paper.

Microgrid configuration

Figure 1 is a schematic of the microgrid configuration considered in this paper. To understand the interaction between grid components and facilitate the development of a mathematical model of the grid's dynamics, the directions of energy flow between grid components have been shown. A two-way energy flow link exists between the MGO and the distribution system operator (DSO) to facilitate energy trading. A similar connection exists between each prosumer and the microgrid for bi-directional flow of energy.

Mathematical model

The objective function is made up of operating cost and emission function components, as detailed in (1). Its operating cost component comprises costs of conventional generation, grid energy purchase minus energy sale, and reserve capacities from both CGs and prosumers' batteries. The emission component comprises CG emissions from power generation and reserve capacity provision.

$$
\text{Min}\ \ \omega 1 * \left(\sum_{t=1}^{T}\sum_{g=1}^{G} C_{c.gen,g}^{t} + \sum_{t=1}^{T}(C_{buy}^{t} * E_{buy}^{t}) \right.
$$
$$
- \sum_{t=1}^{T}(C_{sell}^{t} * E_{sell}^{t}) + \sum_{t=1}^{T}\sum_{g=1}^{G} CR_{c.gen,g}^{t} + \sum_{t=1}^{T}\sum_{p=1}^{P} CR_{Bat,p}^{t} \right)
$$
$$
+ \omega 2 * \left(\sum_{t=1}^{T}\sum_{g=1}^{G}(EM_g * P_{c.gen,g}^{t}) + \sum_{t=1}^{T}\sum_{g=1}^{G}(EMR_g * R_{c.gen,g}^{t}) \right)
$$

$$(1)$$

where:

$$\omega 2 = 1 - \omega 1 \tag{2}$$

$$EMR_g = 0.5 * EM_g \tag{3}$$

$$C_{c.gen,g}^{t} = S_{c.gen,g}^{t} * a_g + b_g * P_{c.gen,g}^{t} \tag{4}$$

$$CR_{c.gen,g}^{t} = 0.75 * (S_{c.gen,g}^{t} * a_g + b_g * R_{c.gen,g}^{t}) \tag{5}$$

$$CR_{Bat,p}^{t} = 0.2 * C_{buy}^{t} * R_{Bat,p}^{t} \tag{6}$$

Constraints (7) through (9) disallow simultaneous purchase/sale of energy from/to the main grid and keep the amount of energy bought/sold per timeslot within permissible limits (Damisa, Nwulu, and Sun 2018).

$$E_{buy}^{t} \leq E_{MiMaTxfGrid}^{max} * S_{buy}^{t} \quad \forall\, t \in [1, T] \tag{7}$$

$$E_{sell}^{t} \leq E_{MiMaTxfGrid}^{max} * S_{Sell}^{t} \quad \forall\, t \in [1, T] \tag{8}$$

$$S_{Buy}^{t} + S_{Sell}^{t} \leq 1 \quad \forall\, t \in [1, T] \tag{9}$$

Constraint (10) is required to make sure that deferrable loads are only served within the period specified by customers (Zhang, Gatsis, and Giannakis 2013), and (11) ensures that their demand is completely met within the period. Interruption frequency of energy supply to these loads is limited by constraint (12). To retain the linearity of the model and guarantee an optimal solution, a new variable, SDL_p^t, is introduced to replace the product of $S_{DL,p}^t * S_{DL,p}^{t+1}$ as seen in constraint (13), and constraints (14) through (17) are used to appropriately limit the new variable (Tazvinga, Xia, and Zhang 2013).

$$S_{DL,p}^{t} = 0 \quad \forall\, end < t < start \tag{10}$$

$$\sum_{t=start}^{end} (S_{DL,p}^t * E_{DL,p}^t) = E_{Dl,p}^{total} \qquad (11)$$

$$\sum_{t=start}^{end-1} (S_{DL,p}^t * S_{DL,p}^{t+1}) \geq \frac{E_{DL,p}^{total}}{E_{DL,p}} - 2 \quad \forall\, p \in [1, P] \qquad (12)$$

$$\sum_{t=start}^{end-1} SDL_p^t \geq \frac{E_{DL,p}^{total}}{E_{DL,p}} - 2 \quad \forall\, p \in [1, P] \qquad (13)$$

$$SDL_p^t \leq S_{DL,p}^t$$
$$\forall\, p \in [1, P]\, \forall\, t \in [start, end - 1] \qquad (14)$$

$$SDL_p^t \leq S_{DL,p}^{t+1}$$
$$\forall\, p \in [1, P]\, \forall\, t \in [start, end - 1] \qquad (15)$$

$$SDL_p^t \geq S_{DL,p}^t + S_{DL,p}^{t+1} - 1$$
$$\forall\, p \in [1, P]\, \forall\, t \in [start, end - 1] \qquad (16)$$

$$SDL_p^t \in [0, 1]$$
$$\forall\, p \in [1, P]\, \forall\, t \in [start, end - 1] \qquad (17)$$

The prosumer's battery dynamics is modelled using Equation (18) and constrained using (19) through (23) (Damisa, Nwulu, and Sun 2018). It should be noted that the variable $E_{BatTxf,p}^t$ can take both positive and negative values. It takes a positive value when the battery is being charged, and a negative value when it is being discharged. Constraint (19) ensures that the battery's content is kept within its capacity limits. Constraints (20) through (22) ensure that the sum of energy to be delivered/absorbed and reserve to be delivered by a battery does not exceed its maximum transfer capacity per timeslot. Constraint (23) represents the limitation on the amount of energy deliverable by the battery due to its efficiency (Zhang, Gatsis, and Giannakis 2013).

$$E_{Bat,p}^t = E_{Bat,p}^{t-1} + n_p(E_{BatTxf,p}^t) - n_p(R_{Bat,p}^t)$$
$$\forall\, p \in [1, P]\, \forall\, t \in [1, T] \qquad (18)$$

$$E_{Bat,p}^{min} < E_{Bat,p}^t < E_{Bat,p}^{max}$$
$$\forall\, p \in [1, P]\, \forall\, t \in [1, T] \qquad (19)$$

$$-E_{BatTxf,p}^{max} \leq E_{BatTxf,p}^t \quad \forall\, p \in [1, P]\, \forall\, t \in [1, T] \qquad (20)$$

$$-E_{BatTxf,p}^t + R_{Bat,p}^t \leq E_{BatTxf,p}^{max}$$
$$\forall\, p \in [1, P]\, \forall\, t \in [1, T] \qquad (21)$$

$$E_{BatTxf,p}^t + R_{Bat,p}^t \leq E_{BatTxf,p}^{max}$$
$$\forall\, p \in [1, P]\, \forall\, t \in [1, T] \qquad (22)$$

$$n_p E_{Bat,p}^{t-1} \geq -E_{BatTxf,p}^t + R_{Bat,p}^t$$
$$\forall\, p \in [1, P]\, \forall\, t \in [1, T] \qquad (23)$$

Constraint (24) is useful for setting the acceptable number of CG interruptions (Damisa, Nwulu, and Sun 2018, 912). Again to retain the linearity of the formulation, another new variable, SG_g^t, is introduced to replace the product of $S_{c.gen,g}^t * S_{c.gen,g}^{t+1}$ as seen in constraint (25), and constraints (26) through (29) are added to appropriately limit this new variable (Tazvinga, Xia, and Zhang 2013). Constraints (30) and (31) make sure that the aggregate of energy and reserve to be provided by a CG is within its capacity limits. Ramp-up and ramp-down restrictions are represented in constraints (32) and (33) (Damisa, Nwulu, and Sun 2018, 913).

$$\sum_{t=1}^{T-1} (S_{c.gen,g}^t * S_{c.gen,g}^{t+1}) = \sum_{t=1}^{T} S_{c.gen,g}^t - 1$$
$$\forall\, g \in [1, G] \qquad (24)$$

$$\sum_{t=1}^{T-1} SG_g^t = \sum_{t=1}^{T} S_{c.gen,g}^t - 1 \quad \forall\, g \in [1, G] \qquad (25)$$

$$SG_g^t \leq S_{c.gen,g}^t \quad \forall\, g \in [1, G]\, \forall\, t \in [1, T - 1] \qquad (26)$$

$$SG_g^t \leq S_{c.gen,g}^{t+1} \quad \forall\, g \in [1, G]\, \forall\, t \in [1, T - 1] \qquad (27)$$

$$SG_g^t \geq S_{c.gen,g}^t + S_{c.gen,g}^{t+1} - 1$$
$$\forall\, g \in [1, G]\, \forall\, t \in [1, T - 1] \qquad (28)$$

$$SG_g^t \in [0, 1] \quad \forall\, g \in [1, G]\, \forall\, t \in [1, T - 1] \qquad (29)$$

$$S_{c.gen,g}^t * P_{c.gen,g}^{min} \leq P_{c.gen,g}^t + R_{c.gen,g}^t$$
$$\leq P_{c.gen,g}^{max} * S_{c.gen,g}^t \qquad (30)$$
$$\forall\, g \in [1, G]\, \forall\, t \in [1, T]$$

$$S_{c.gen,g}^t * P_{c.gen,g}^{min} \leq P_{c.gen,g}^t \leq P_{c.gen,g}^{max} * S_{c.gen,g}^t$$
$$\forall\, g \in [1, G]\, \forall\, t \in [1, T] \qquad (31)$$

$$P_{c.gen,g}^t + R_{c.gen,g}^t - P_{c.gen,g}^{t-1} - R_{c.gen,g}^{t-1} \leq r * P_{c.gen,g}^{max}$$
$$\forall\, g \in [1, G]\, \forall\, t \in (1, T] \qquad (32)$$

$$P_{c.gen,g}^{t-1} + R_{c.gen,g}^{t-1} - P_{c.gen,g}^t - R_{c.gen,g}^t \leq r * P_{c.gen,g}^{max}$$
$$\forall\, g \in [1, G]\, \forall\, t \in (1, T] \qquad (33)$$

The required reserve capacity is assumed to be the sum of maximum possible downward deviation from generation forecasts and maximum possible upward deviation from demand forecasts, as shown in constraint (34). CGs, prosumers' batteries and deferrable loads are expected to provide this capacity, as represented by constraint (35)

(Damisa, Nwulu, and Sun 2018, 913).

$$E_R^t = DevE_{Wind}^{-,t} + DevE_{Solar}^{-,t} + DevE_{Demand}^{+,t}$$
$$+ \sum_p^P DevE_{\Pr oGen,p}^{-,t} + \sum_p^P DevE_{\Pr oDem,p}^{+,t} \quad (34)$$
$$\forall t \in [1, T]$$

$$\sum_{g=1}^{G} R_{c.gen,g}^t + \sum_{p=1}^{P} R_{Bat,p}^t + \sum_{p=1}^{P} E_{DL,p} * S_{DL,p}^t$$
$$\geq E_R^t \quad \forall t \in [1, T] \quad (35)$$

Two energy balance equations are necessary for this grid configuration; one for prosumers and the other for the microgrid as a whole entity (Damisa, Nwulu, and Sun 2018). Equation (36) ensures energy balance is the prosumer's premises, and Equation (37) maintains energy balance in the microgrid.

$$E_{ProGen, p}^t - E_{GridPro,p}^t - E_{NDL,p}^t - E_{BatTxf,p}^t - E_{DL,p} * S_{DL,p}^t$$
$$= 0 \quad \forall p \in [1, P] \, \forall t \in [1, T]$$
$$(36)$$

$$E_{Solar}^t + E_{Wind}^t + \sum_{g=1}^{G} E_{c.gen,g}^t + E_{buy}^t$$
$$+ \sum_{p=1}^{P} E_{GridPro,p}^t$$
$$= E_{sell}^t + E_{OtherLoad}^t \quad \forall t \in [1, T] \quad (37)$$

Simulation set-up

The modified microgrid test system, as well as grid components and their sizes, used as case study in this work, is similar to that used in Damisa, Nwulu, and Sun (2018, 914). The average solar irradiance forecast is taken from Hung, Mithulananthan, and Lee (2014). Microgrid and commercial and industrial prosumer demands are obtained using the percentage of peak demand values in Papathanassiou, Hatziargyriou, and Strunz (2005) and Force (1979). 400 kW, 150 kW and 50 kW are assumed to be the peak demands for the microgrid, industrial and commercial prosumers respectively. Data and details regarding calculation of maximum

deviation from wind, solar and demand forecasts can be found in Damisa, Nwulu, and Sun (2018, 913). The average wind speed forecast and cost of energy from the main grid are taken from Zakariazadeh, Jadid, and Siano (2014b), and the cost of selling = cost of buying – 0.01$, to avoid energy arbitrage Pal et al. (2016). CG parameters used are given in Table 1, and $r = 0.2$. Network power flow equations are not considered in this work. The results presented below were obtained using Advanced Interactive Multidimensional Modelling Systems (AIMMS) (Nwulu and Xia 2015b), on a personal computer with processor: Intel(R) Pentium (R) Dual CPU T2390 @ 1.86 GHz 1.87 GHz.

Results

In this work, three different cases are considered, viz. the minimization of operating cost alone ($w1 = 1$), minimization of emissions alone ($w1 = 0$), and minimization of both operating cost and emissions with each carrying equal weight in the objective function ($w1 = 0.5$). It should be noted that the second weight factor, $w2 = 1 - w1$. Tables 2 and 3 present the total energy and reserve schedules, respectively, for each of the three scenarios considered. Table 4 gives the total operating cost and emission per case. Figures 2–4 compare the scheduled energy purchase, scheduled energy sale and scheduled CG generation, respectively, for the three scenarios. In Figures 5–7, comparisons of the scheduled reserve from prosumer batteries, CGs and DLs, respectively, for the three scenarios are shown.

Discussion of results

Notice in Table 2 that the total energy generation scheduled for CG 1 far outweighs that for CG 2. This is because CG 1 has a relatively lower cost of generation. Reserves from CG 1 are also more than those from CG 2 except in Case 2, because more CG generation is used in Case 2 than in the two other cases, since emission is not taken into consideration; CG 1, being the cheaper producer of power, is utilized more (sometimes at its peak), so CG 2 which is used less has more available reserve capacity. The total energy scheduled to be bought from the main grid is also significantly lower in Case 2 because more CG generation is utilized since their emission production is not a concern.

It is interesting to note from Table 4 that similar values were obtained for both total operating cost and

Table 1: CG parameters (Zakariazadeh, Jadid, and Siano 2014a, 2014b).

CG	Min. power (kW)	Max. power (kW)	a_g ($)	b_g ($/kWh)	CO_2 emission (kg/kWh)
1	30	300	0.5	0.053	0.74
2	40	400	0.8	0.068	0.89

Table 2: Energy schedule.

Case	Total energy bought (kWh)	Total energy sold (kWh)	Total CG 1 output (kWh)	Total CG 2 output (kWh)
Case 1 $w1 = 0.5$	4529.04	0.00	2847.05	80.00
Case 2 $w1 = 1.0$	2339.86	964.51	4353.59	1654.05
Case 3 $w1 = 0.0$	4529.04	0.00	2847.05	80.00

Table 3: Reserve schedule.

Case	Total CG 1 reserve (kWh)	Total CG 2 reserve (kWh)	Total ind. pro. battery reserve (kWh)	Total comm. pro. battery reserve (kWh)
Case 1 $w1 = 0.5$	734.48	60.40	98.59	54.21
Case 2 $w1 = 1.0$	360.16	507.84	49.69	30.00
Case 3 $w1 = 0.0$	734.48	60.40	108.42	44.38

Table 4: Total operating cost and emissions.

Case	Total operating cost ($)	Total emissions (kg)
Case 1 $w1 = 0.5$	640	2477
Case 2 $w1 = 1.0$	277	5053
Case 3 $w1 = 0.0$	640	2477

emissions in Cases 1 and 3. Hence, in this grid configuration, minimizing the objective function with equal weights assigned to operating cost and emissions (Case 1) yields the same results as minimizing emissions alone (Case 3). Also, looking closely at Figures 2–7, the schedules generated for these two cases are similar, for the most part. The similarities are found mainly in

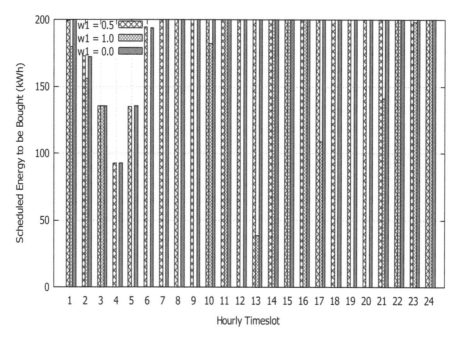

Figure 2: Scheduled energy to be bought for each of the three cases.

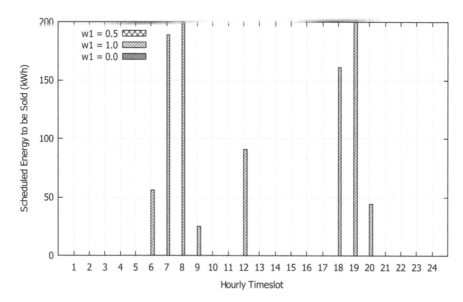

Figure 3: Scheduled energy to be sold for each of the three cases.

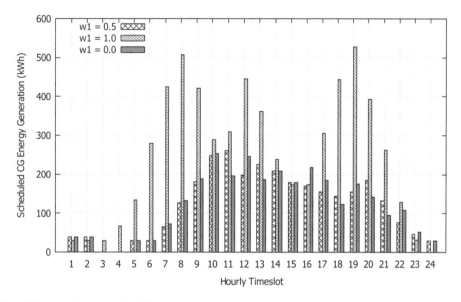

Figure 4: Scheduled CG generation for each of the three cases.

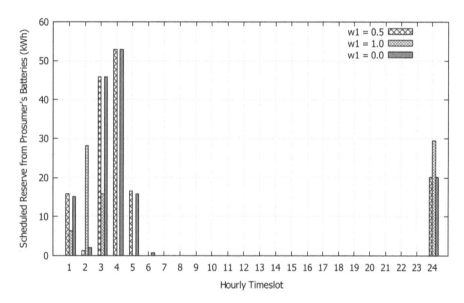

Figure 5: Scheduled reserve from prosumers' batteries for each of the three cases.

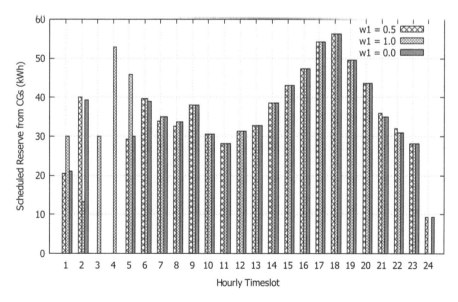

Figure 6: Scheduled reserve from CGs for each of the three cases.

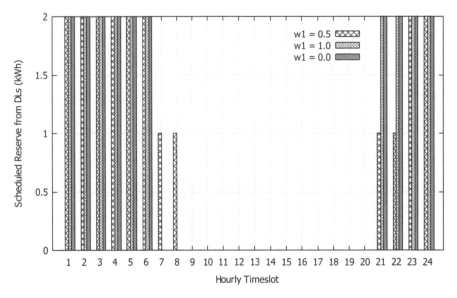

Figure 7: Scheduled reserve from DLs for each of the three cases.

the schedule for energy to be bought, CG energy generation and reserve provision, and reserve from prosumers' batteries. However, a significant dissimilarity is observed in the schedule for DLs. The observed similarity in these two cases may be as a result of the relatively higher value of emissions in comparison with operating cost. From the foregoing, it can be concluded that equal weights assigned to operating cost and emissions, as in Case 1, does not necessarily imply that both are effectively equally weighted.

In Cases 1 and 3 (where emission is taken into account), both CGs are scheduled to be turned off in timeslots 3 and 4, during which prosumers' batteries are scheduled to assist deferrable loads in providing the required reserve capacity. In Case 2, on the other hand, CGs are scheduled to be on throughout the scheduling horizon, except in timeslot 24. Also, Table 3 shows that the total reserve capacity expected from prosumers' batteries is more in Cases 1 and 3 than in Case 2. Hence, besides reducing CG emissions, the inclusion of an emission function in the objective function of an operational dispatch formulation, is beneficial to customers.

In timeslots seven, eight, 18 and 19, the cost of selling energy to the main grid is high, and so much profit can be made for the microgrid if energy is sold during these periods, however, due to the limitation placed on CG generation as a result of trying to minimize emissions, no energy is scheduled for sale in Cases 1 and 3. On the other hand, much profit is earned from energy sale in Case 2, as seen in Figure 3. So, though the amount of emission is lower in Cases 1 and 3 than in Case 2, the profit earned from selling energy during periods of high energy prices is higher in Case 2.

The following are policy recommendations:

• Energy consumers should be incentivized to offer significant portions of their loads for direct load control as this plays a key role in curbing emissions.

• Prosumer behind-the-meter resources such as battery storage facilities can provide grid ancillary services thereby relieving CG utilization, and hence reducing emissions; investments in these facilities should therefore be motivated.

Conclusion

A mathematical model of the economic and emission dispatch problem of a microgrid was presented in this paper. The model was formulated as a mixed-integer multi-objective optimization problem whose solution minimizes the operating cost of the microgrid as well as CG emissions. The weighting method of tackling multi-objective optimization problems was employed in this study, and three weight combinations were investigated. While in the first weight combination, equal weights were assigned to both generation cost and emission, the second and third combinations minimized only generation cost and only emissions, respectively. Results from investigations carried out on a modified microgrid test system, and comparisons of schedules generated for the different weight combinations, show that the consideration of emission in the operational planning of a microgrid could be of advantage to customers as more of their onsite resources may be utilized to supply grid support services. On the other hand, restriction may be placed on the profit that can be made from selling energy during periods of high energy price, when emission reduction is an objective of the operational dispatch model.

The development of a stochastic variant of the formulation presented in this paper is a possible way to extend the work. Emissions resulting from the production and transportation of prosumers' batteries may also be included in the formulation.

Disclosure statement

No potential conflict of interest was reported by the authors.

Funding
Yanxia Sun, the third author of this paper, acknowledges the support received from the South African National Research Foundation Grants (No. 112108 and 112142), South African National Research Foundation Incentive Grant (No. 95687), and a research grant from URC of University of Johannesburg.

References
Ahn, C., and H. Peng. 2013. "Decentralized and Real-time Power Dispatch Control for an Islanded Microgrid Supported by Distributed Power Sources." *Energies* 6 (12): 6439–6454. doi:10.3390/en6126439.

Bestehorn, M., and T. Borsche. 2014. "Balancing Power Consumption and Production in Smart Grids." Paper presented at innovative smart grid technologies conference Europe (ISGT-Europe), IEEE PES, October 12, 1–6.

Chen, S., N. B. Shroff, and P. Sinha. 2013. "Heterogeneous Delay Tolerant Task Scheduling and Energy Management in the Smart Grid with Renewable Energy." *IEEE Journal on Selected Areas in Communications* 31 (7): 1258–1267. doi:10.1109/JSAC.2013.130709.

Couture, T., G. Barbose, D. Jacobs, G. Parkinson, E. Chessin, A. Belden, H. Wilson, H. Barrett, and W. Rickerson. 2014. *Residential Prosumers: Drivers and Policy Options (Re-prosumers).* Berkeley, CA: Meister Consultants Group; Lawrence Berkeley National Lab. (LBNL).

Damisa, U., N. I. Nwulu, and Y. Sun. 2018. "Microgrid Energy and Reserve Management Incorporating Prosumer Behind-the-Meter Resources." *IET Renewable Power Generation* 12 (8): 910–919. doi:10.1049/iet-rpg.2017.0659.

Farimani, F. D., and H. R. Mashhadi. 2013. "Effects of Demand Dispatch on Operation of Smart Hybrid Energy Systems." Paper presented at power systems conference, Iran, November.

Farimani, F. D., and H. R. Mashhadi. 2015. "Modeling and Implementation of Demand Dispatch Approach in a Smart Micro-Grid." In *Integral Methods in Science and Engineering*, edited by C. Constanda and A. Kirsch, 129–141. Cham: Birkhäuser, Springer International Publishing. doi:https://doi.org/10.1007/978-3-319-16727-5_12.

Force, R. T. 1979. "IEEE Reliability Test System." *IEEE Transactions on PAS* 98 (6): 2047–2054. doi:10.1109/TPAS.1979.319398.

He, P., M. Li, L. Zhao, B. Venkatesh, and H. Li. 2018. "Water-filling Exact Solutions for Load Balancing of Smart Power Grid Systems." *IEEE Transactions on Smart Grid* 9 (2): 1397–1407. doi:10.1109/TSG.2016.2590147.

Huang, W.-T., K.-C. Yao, and C.-C. Wu. 2014. "Using the Direct Search Method for Optimal Dispatch of Distributed Generation in a Medium-voltage Microgrid." *Energies* 7 (12): 8355–8373. https://doi.org/10.3390/en7128355.

Hung, D. Q., N. Mithulananthan, and K. Y. Lee. 2014. "Determining PV Penetration for Distribution Systems with Time-varying Load Models." *IEEE Transactions on Power Systems* 29 (6): 3048–3057. doi:10.1109/TPWRS.2014.2314133.

Nguyen, T. A., and M. L. Crow. 2016. "Stochastic Optimization of Renewable-based Microgrid Operation Incorporating Battery Operating Cost." *IEEE Transactions on Power Systems* 31 (3): 2289–2296. doi:10.1109/TPWRS.2015.2455491.

Nwulu, N. I., and X. Xia. 2015b. "Implementing a Model Predictive Control Strategy on the Dynamic Economic Emission Dispatch Problem with Game Theory Based Demand Response Programs." *Energy* 91: 404–419. https://doi.org/10.1016/j.energy.2015.08.042.

Nwulu, N. I., and X. Xia. 2015a. "Multi-objective Dynamic Economic Emission Dispatch of Electric Power Generation Integrated with Game Theory Based Demand Response Programs." *Energy Conversion and Management* 89: 963–974. https://doi.org/10.1016/j.enconman.2014.11.001.

Pal, R., C. Chelmis, M. Frincu, and V. Prasanna. 2016. "MATCH for the Prosumer Smart Grid The Algorithmics of Real-Time Power Balance." *IEEE Transactions on Parallel and Distributed Systems* 27 (12): 3532–3546. doi:10.1109/TPDS.2016.2544316.

Papathanassiou, S., N. Hatziargyriou, and K. Strunz. 2005. "A Benchmark Low Voltage Microgrid Network." Paper presented at the CIGRE symposium, Athens, April 13, 1–8.

Salinas, S., M. Li, P. Li, and Y. Fu. 2013. "Dynamic Energy Management for the Smart Grid with Distributed Energy Resources." *IEEE Transactions on Smart Grid* 4 (4): 2139–2151. doi:10.1109/TSG.2013.2265556.

Su, H.-I., and A. El Gamal. 2013. "Modeling and Analysis of the Role of Energy Storage for Renewable Integration: Power Balancing." *IEEE Transactions on Power Systems* 28 (4): 4109–4117. doi:10.1109/TPWRS.2013.2266667.

Tazvinga, H., X. Xia, and J. Zhang. 2013. "Minimum Cost Solution of Photovoltaic–Diesel–battery Hybrid Power Systems for Remote Consumers." *Solar Energy* 96: 292–299. https://doi.org/10.1016/j.solener.2013.07.030.

Zafar, R., A. Mahmood, S. Razzaq, W. Ali, U. Naeem, and K. Shehzad. 2018. "Prosumer Based Energy Management and Sharing in Smart Grid." *Renewable and Sustainable Energy Reviews* 82 (1): 1675–1684. https://doi.org/10.1016/j.rser.2017.07.018.

Zakariazadeh, A., S. Jadid, and P. Siano. 2014a. "Economic-environmental Energy and Reserve Scheduling of Smart Distribution Systems: A Multiobjective Mathematical Programming Approach." *Energy Conversion and Management* 78: 151–164. https://doi.org/10.1016/j.enconman.2013.10.051.

Zakariazadeh, A., S. Jadid, and P. Siano. 2014b. "Smart Microgrid Energy and Reserve Scheduling with Demand Response Using Stochastic Optimization." *International Journal of Electrical Power & Energy Systems* 63: 523–533. https://doi.org/10.1016/j.ijepes.2014.06.037.

Zhang, Y., N. Gatsis, and G. B. Giannakis. 2013. "Robust Energy Management for Microgrids with High-Penetration Renewables." *IEEE Transactions on Sustainable Energy* 4 (4): 944–953. doi:10.1109/TSTE.2013.2255135.

A simplified control scheme for electric vehicle-power grid circuit with DC distribution and battery storage systems

Kabeya Musasa, Musole Innocent Muheme, Nnamdi Ikechi Nwulu and Mammo Muchie ⓘ

The direct current (DC) system is becoming the major trend for future internal power grid of electric vehicles (EVs). Since DC power grid system has a different nature to conventional alternating current (AC) grid system, appropriate design of the controller for EV-grid circuit is mandatory. In this paper, an EV employing a pure DC grid circuit with battery storage system (BSS) is considered as a study case. To enable a more efficient use of BSS, a flyback DC-DC converter for batteries charger/or discharger strategy is selected, which satisfies the power flows requirements. The dynamic and control performances of the combined system, i.e. 'BSS-flyback DC-DC converter connected to a DC motor', is investigated in terms of voltage/current signal fluctuations. The small-signal based control method is used, which limits the small-signal variations to about zero. To verify the effectiveness of the control strategy, several simulations are done using Matlab. The simulation results illustrate the performances obtained.

Introduction

The negative environmental impacts of the existing transportation system have made it mandatory to search for alternative solutions to prevent vast environmental pollution. The introduction of electric vehicles (EV) has proved that they are very environmentally friendly and emit fewer pollutants compared to vehicles with petrol/diesel engines (Sankara and Seyezhai 2016). Globally, the concept of electrically powered road vehicles is developing very fast. It has a good chance of becoming a significant part of the transportation infrastructure in the near future.

Among the components that are actually needed in an EV-power grid circuit are: an electric drive or electric motor, a power converter with controller scheme, a battery storage system (BSS), etc. Figure 1 shows an example of an electric power grid scheme for an EV (Biradar, Patil, and Ullegaddi 1998). The energy that flows into the EV-power grid comes directly from the BSS. In Sankara and Seyezhai (2016) and Biradar, Patil, and Ullegaddi (1998), the BSS was combined with the photovoltaic (PV) solar system to power up the EV-power grid. The operating principle of a PV solar for EV can be found in Sankara and Seyezhai (2016).

They are three different types of EV models that are being developed for use in the transportation sector to reduce the environmental pollution problem. These include: the hybrid electric vehicles (HEVs), the plug-in hybrid electric vehicles (PHEVs), and the pure battery electric vehicles (BEVs) (Young et al. 2013; Momoh and Omoigui 2009). The HEVs are very popular nowadays. But pure BEVs are the best, most-highly developed models.

This paper analyzes the control circuit of a pure BEV model. The BSS is a fundamental component of BEV; it powers up the EV-electric motor through a power electronic converter. The batteries for EVs are different from those used in small electronic devices like laptops and cell phones. The batteries for an EV require handling high power or high energy capacity and they must also be very compact or small in size and weight for them to be installed within a limited space. Extensive research is being conducted to develop advanced battery technologies that are suitable for EVs all over the world (Young et al. 2013).

Furthermore, in order to enable a more efficient use of energy from a BSS, a dedicated power electronic converter for batteries charger/or discharger strategy must be selected, which can satisfy the power flow requirements. In this paper an isolated flyback DC-DC converter is selected to interface the BSS to a DC motor. In Anwar et al. (2016), the dual active bridge (DAB) DC-DC converter was used to interface the BSS to a DC motor. The DAB DC-DC converter consists of a high-frequency inverter, a high-frequency transformer, and a high-frequency active bridge-rectifier. This converter topology enables a bidirectional power flow between the BSS and the DC motor.

The bidirectional operating condition of the power converter is necessary in EV-power grid circuits. This is because in the armature winding of the DC motor, a back-electromagnetic force (emf) is generally produced by the rotation of armature conductors in the presence of field (or flux). The energy extracted from the back-emf can be useful to recharge the BSS. But such scheme will require a very complex EV-power grid circuit.

Instead of using the DAB DC-DC converter topology to interface the BSS and the DC motor, a single active bridge (SAB) DC-DC converter topology can also be used. The SAB DC-DC converter topology consists of a high-frequency inverter, a high-frequency transformer, and a high-frequency diode bridge-rectifier. Actually, the SAB DC-DC converter topology does not permit a bidirectional power flow between BSS and DC motor because of the passive diode-bridge rectifier. In Pany, Singh, and Tripathi (2011), the bidirectional DC-DC

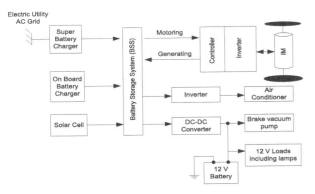

Figure 1: Sample of a power grid scheme for an EV (Biradar, Patil, and Ullegaddi 1998).

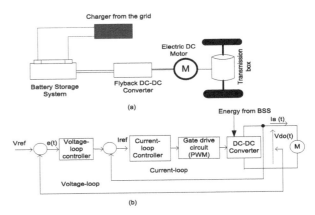

Figure 2: Simplified equivalent block diagram of an EV: (a) power grid circuit; (b) control scheme installed across the flyback DC-DC converter.

buck converter was selected to interface the BSS and the DC motor. A three-phase full-bridge DC-DC converter was considered in Kumar and Gaur (2014). And the buck-boost DC-DC converter was used in Albiol-Tendillo et al. (2012). In all these papers, the DC-DC converters were properly selected, but no specific control strategy was elaborated.

The types of electric drives or motors used for EVs include the brushed DC motor, brushless DC motor, AC induction motor, permanent magnet synchronous motor (PMSM) and switched reluctance motor (SRM) (Huang, Li, and Chen 2010). The fact that the power output of BSS is DC power, a DC motor is selected in this paper to drive the EV wheels. There are three types of brushed DC motor with field windings, which are the series, shunt and separately excited windings (Huang, Li, and Chen 2010). The topology of DC motors for EV with magnet ring and stator winding are presented in Lovatt, Ramsden, and Mecrow (1998).

In this paper, the dynamic and control strategy of the combined system, i.e. "BSS-flyback DC-DC converter connected to a DC motor", is investigated in terms of voltage/current signal fluctuations. The dynamic model of the flyback DC-DC converter is also derived. The small-signal based control method is used, which limits the small-signal variations to about zero. The effectiveness of the control strategy which is provided by the flyback DC-DC converter is tested by using an arbitrary EV-data. To verify the effectiveness of the control strategy several simulations are done using Matlab.

EV-power grid: simplified equivalent block diagram
Figure 2(a) presents a simplified equivalent block diagram of the EV-power grid discussed in this paper. The EV-power grid includes a control scheme installed across the flyback DC-DC converter, depicted by Figure 2(b), and an electric DC motor connected to two wheels via a transmission box. The DC motor is powered by a BSS via a flyback DC-DC converter. The batteries can be interconnected either in series to increase the voltage or in series-parallel to increase both the voltage and the current or power. The batteries are recharged from the mains electrical network.

The typical EV-power grid circuit with a single closed-loop control scheme was discussed in Santos et al. (2007).

The flyback DC-DC converter controls the power delivered to the DC motor through variation of the duty ratio of the converter switches. Based on Figure 2(b), the gate drive circuit, which consists of a pulse-width modulation (PWM) scheme, commands the state of the switches of the flyback DC-DC converter. Depending on the performance to achieve, the current controller or compensator in the current-loop system can either be the proportional–integral (PI) or the proportional–integral–derivative (PID) controller. A filter can also be required at the output of the DC-DC converter to reduce the ripple current in the DC motor.

DC motor: Principle of operation
In a DC motor, the field flux φ_f is established by the stator, either by means of permanent magnets, where φ_f =constant, or by means of a field winding, where φ_f is controlled by i_f; as given by eq. (1) (Mohan, Undeland, and Robbins 2003), with k_f being the field constant of proportionality, and $i_f(t)$ the field current.

$$\varphi_f = k_f i_f \tag{1}$$

The armature winding on the rotor side carries the electrical power from BSS via the split ring commutator or brushes. The electromagnetic torque is produced by the interaction of $\varphi_f(t)$ and $i_a(t)$, as eq. (2) (Pany, Singh, and Tripathi 2011), where $i_a(t)$ is the armature winding current, and k_t the torque constant of the rotor.

$$T_{em} = k_t \varphi_f i_a \tag{2}$$

The armature current $i_a(t)$ is determined by a controllable $v_{do}(t)$ (i.e. the output voltage of the flyback DC-DC converter), by the armature winding resistance r_a, by the armature winding inductance L_a, and by the induced back-emf $e_a(t)$; this is derived in Laplace domain as eq. (3).

$$i_a(s) = \frac{v_{do}(s) - e_a(s)}{r_a + L_a s} \tag{3}$$

The back-emf is produced by the rotation of armature conductors at speed ω_m in the presence of $\varphi_f(t)$, as eq. (4)

(Veilleux and Lehn 2014), where k_e is the voltage constant of the electric motor.

$$e_a(t) = k_e \varphi_f \omega_m \qquad (4)$$

Based on Figure 2(a), the DC motor is driving two wheels through a transmission box at a speed of ω_m. During the braking operation: if $|v_{do}(t)| < |e_a(t)|$, $|i_a(t)|$ will reverse in direction, also T_{em} defined by eq. (2) will reverse in direction. As a result, the kinetic energy associated with the motor inertia is converted into electrical energy by the DC machine, which acts as a DC generator. This energy can either be dissipated in a resistor or absorbed by BSS. As ω_m decreases (the motor slows down) and $e_a(t)$ decreases in magnitude.

DC motor with magnet ring
The permanent magnets on the stator side produce a constant φ_f, and thus eq. (2), (3), and (4) will not depend on the field current i_f. As a result, eq. (2), (3), and (4) can, respectively, take the form in eq. (5):

$$T_{em} = k_T i_a; \quad V_{do} = E_a + r_a i_a; \quad \text{and } e_a(t) = k_{PM} \omega_m \qquad (5)$$

where $k_T = k_t \varphi_f$ and $k_{PM} = k_e \varphi_f$. From eq. (5), one can derive the expression of ω_m in terms of T_{em} given different values of $v_{do}(t)$: i.e. the torque-speed characteristics, eq. (6):

$$\omega_m = \frac{1}{k_{PM}}\left(v_{do}(t) - \frac{r_a}{k_T} T_{em}\right) \qquad (6)$$

Based on eq. (6), it is noted that the rotating speed of wheels for an EV with permanent magnet DC motor is controlled by controlling the output voltage of the flyback DC-DC converter, $v_{do}(t)$.

DC motor with separately excited winding
Permanent magnet-based DC motors have maximum speed limitation. To overcome this limitation, the field winding on the stator side is excited by a separately controlled DC source. Using eq. (1), $\varphi_f(t)$ is controlled through $i_f(t)$, and thus eq. (6) can take the form in eq. (7):

$$\omega_m = \frac{1}{k_e \varphi_f}\left(v_{do}(t) - \frac{r_a}{k_t \varphi_f} T_{em}\right) \qquad (7)$$

From eq. (7), it is noted that the rotating speed of the wheels in an EV employing a DC motor with separately excited winding, is controlled by controlling both the output voltage of the flyback DC-DC converter $v_{do}(t)$ and the field winding $\varphi_f(t)$.

Effect of armature voltage and current ripples
Based on Figure 2, the output voltage of the flyback DC-DC converter may contain small-signal variations, which can lead further to a ripple in the armature current. This will result in a high form factor, high losses in the DC motor, and hence, low efficiency of the DC motor. The form factor is defined as the ratio between the armature

current (rms) and the armature current (average). If i_a is a pure DC signal, then the form factor is unit. The form factor increases when the signal i_a contains small variations or deviates from a pure DC signal.

In addition, from eq. (2), $T_{em}(t)$ depends on $i_a(t)$; thus, a ripple in $i_a(t)$ will result in the torque pulsation, which will lead further to a ripple in ω_m.

EV servo drive: Equivalent model
The proposed EV-controller scheme or -regulator block diagram to compensate the voltage/current ripple is depicted by Figure 3(a) (Musasa, Gitau, and Bansal 2015a, 2015b); the different transfer functions or equivalent model of the flyback DC-DC converter are derived further in eq. (8), (9), and (10).

The inputs to the regulator are the voltage and current ripples. These ripples originate from the BSS or the DC motor, or any other phenomena. The symbol (^) stands for small-signal variations or ripple. The topology of the isolated flyback DC-DC converter is shown in Figure 3 (b). The goal of this converter is to regulate and smooth the net power injected to the DC motor. Only a single active switch is installed in this converter topology; the overvoltage spike is applied frequently across the single switch S_1 at each turn off. The peak value of this overvoltage depends upon the switching time.

In practice, the single-active-switch flyback DC-DC converter requires a snubber circuit to limit the voltage spike across the switch S_1. The double-active-switch flyback DC-DC converter, depicted by Figure 3(c), is another option that can be considered to limit the voltage spike across switch S_1.

Flyback DC-DC converter: Equivalent models
In Figure 3(b), the flyback transformer has an equivalent magnetizing inductance L_{eq}, and turn ratio 1: n; switch

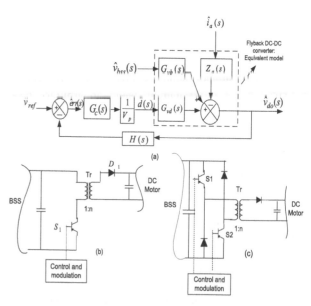

Figure 3: (a) Block diagram of the control scheme. It enables compensating the small current/voltage signal variations; (b) the single active-switch flyback DC-DC converter; and (c) the double active-switch flyback DC-DC converter; this is an alternative topology of the single-switch flyback converter which limits the voltage spike.

S_1 has on-resistance R_{on}; $err(t)$ is the steady-state error; the transformer leakage inductances and the switching loss are negligible.

The transfer functions in Figure 3(a) are derived in eq. (8), (9), and (10); the derivation method of these transfer functions can be found in Erickson (2001) and Musasa, Gitau, and Bansal (2015b), where $G_{vb}(s)$ represents the effect of BSS-output voltage variations on $v_{do}(s)$; $G_{vd}(s)$ represents the effects of duty ratio variations on $v_{do}(s)$; $Z_a(s)$ represents the effects of armature current variations on $v_{do}(s)$; V_{do} and V_{bss} are, respectively, the steady-state voltage input to the DC motor and output terminal of the BSS.

$$G_{vb}(s) = \frac{\hat{v}_{do}}{\hat{v}_{bss}}$$
$$= \left(-\frac{Dn}{1-D}\right) \frac{1}{1 + s\left(\frac{L_{eq}n^2}{(1-D)^2 r_a}\right) + s^2\left(\frac{L_{eq}Cn^2}{(1-D)^2}\right)} \quad (8)$$

$$G_{vd}(s) = \frac{\hat{v}_{do}}{\hat{d}(s)}$$
$$= \left(\frac{V_{do} - V_{bss}}{1-D}\right) \frac{1 - s\left(\frac{L_{eq}I_a}{(1-D)(V_{bss} - V_{do})}\right)}{1 + s\left(\frac{L_{eq}n^2}{(1-D)^2 r_a}\right) + s^2\left(\frac{L_{eq}Cn^2}{(1-D)^2}\right)} \quad (9)$$

$$Z_a(s) = \frac{\hat{v}_{do}}{\hat{i}_a} = \frac{s\frac{L_{eq}n^2}{(1-D)^2}}{1 + s\frac{L_{eq}n^2}{r_a(1-D)^2} + s^2\frac{L_{eq}Cn^2}{(1-D)^2}} \quad (10)$$

The regulator system, Figure 3(a), is designed in order to adjust automatically the duty cycle as necessary to obtain a desired $v_{do}(s)$ regardless of disturbances in EV-power grid circuit parameters.

The inductor and capacitor of the flyback DC-DC converter can be sized using eq. (11) and (12), respectively, where Δi_{pk-pk} and Δv_{pk-pk} are respectively specified peak-to-peak inductor current and capacitor voltage ripples. The steady-state component of the duty cycle can be determined using eq. (13). However, these values can be adjusted such that a given or desired performance is achieved.

$$L_{eq} = \frac{V_{bss} - I_{pcc}R_{on}}{\Delta i_{pk-pk}} DT_s \quad (11)$$

$$C = \frac{V_{pcc}DT_s}{R_{pcc}\Delta v_{pk-pk}} \quad (12)$$

$$D \approx \frac{V_{do}}{V_{do} + V_{bss}} \quad (13)$$

Design of compensator transfer function

It is desired to achieve a constant voltage $v_{do}(t) = V_{do}$, in spite of disturbances in EV-parameters. Technically, it may be impossible to achieve such a condition without the use of a compensator in the feedback control-loop system. The compensator or controller $G_c(s)$ is generally designed to attain adequate phase margin and good rejection of expected disturbances, or to reduce the influence of $\hat{v}_{do}(t)$ on the armature current $i_a(t)$.

The Proportional Integral Derivative (PID) compensator or the combined lead-lag compensator system can lead to a very small steady-state error and better rejection of disturbances. Based on the regulator block diagram of Figure 3a, the closed-loop transfer functions are derived in eq. (14), (15), and (16), where $T_v(s)$ is the voltage loop gain given by eq. (17); $H(s)$ is the voltage sensor gain; and V_P is the peak magnitude of the PWM triangular signal.

$$G_{v\_vref}(s) = \frac{\hat{v}_{do}(s)}{\hat{v}_{ref}(s)} = \frac{1}{H(s)}\frac{T_v(s)}{1 + T_v(s)} \quad (14)$$

$$G_{v\_vbss}(s) = \frac{\hat{v}_{do}(s)}{\hat{v}_{bss}(s)} = \frac{G_{vb}(s)}{1 + T_v(s)} \quad (15)$$

$$G_{v\_i}(s) = \frac{\hat{v}_{do}(s)}{\hat{i}_a(s)} = -\frac{Z_{pcc}(s)}{1 + T_v(s)} \quad (16)$$

$$T_v(s) = H(s)G_c(s)G_{vd}(s)/V_P \quad (17)$$

The feedback control-loop system of Figure 3(a) is stable when the phase margin of $T_v(s)$ is positive; the phase margin is defined by eq. (18), where f_c is the crossover frequency.

$$\phi_m = 180° + \angle T_v(j2\pi f_c) \quad (18)$$

Thus, to ensure a stable feedback loop system the characteristic in eq. (19) must be achieved:

$$\angle T_v(j2\pi f_c) > -180° \quad (19)$$

In addition, ϕ_m must be sufficiently large to limit the amount of overshoot and ringing; also, $T_v(s)$ must be sufficiently large in magnitude to limit the magnitude of $Z_a(s)$ and $G_{vb}(s)$ based on eq. (15) and (16). Furthermore, the transient response time can be shortened by increasing the feedback loop crossover frequency. All these specifications can be achieved by a proper design of $G_c(s)$ in eq. (17).

The PID compensator has a transfer function given by eq. (20) (Erickson 2001),

$$G_c(s) = K_c \frac{\left(1 + \frac{2\pi f_L}{s}\right)\left(1 + \frac{s}{2\pi f_z}\right)}{\left(1 + \frac{s}{2\pi f_{p1}}\right)\left(1 + \frac{s}{2\pi f_{p2}}\right)} \quad (20)$$

where f_L is the inverted zero which improves the low-frequency regulation; K_c is the constant gain of $G_c(s)$; f_z is the zero frequency which adds a phase lead at the vicinity of the crossover frequency; and f_{p1} and f_{p2} are the high-frequency poles which prevent the switching ripple from disrupting the operation of the PWM.

PID controller: performance analysis

Based on Figure 3, let's consider a BSS with a nominal output voltage $v_{bss} \approx V_{bss} = 40$ V DC ($\hat{v}_{bss} \approx 0$). It is desired to maintain a regulated voltage $V_{do} = 100$ V DC at the input of the DC motor. By assuming a small armature resistance, $r_a = 2\Omega$; the steady-state armature current is obtained as $I_a = 50$A. A voltage reference of 5 V is selected, and thus $H(s) = V_{ref}/v_{do} = 0.05$ ($err(s) \approx 0$). The steady-state value of the duty cycle is obtained using eq. (13), $D \approx 0.7$; the switching frequency is given $f_s = 1/T_s = 3.5$kHz. The network components of the flyback converter can be sized using eq. (11) and (12) or can be selected from the 'market list of available components': (assuming that: $L_{eq}n^2 \approx 1H$ and $Cn^2 \approx 0.5F$). The voltage loop gain $T_v(s)$ is therefore obtained by substitution of eq. (9) and (20) into (17).

The characteristic of the uncompensated voltage loop gain $T_u(s)$, with unity compensator gain $G_c(s) = 1$, is sketched in Figure 4. It can be seen that the uncompensated loop gain has a very small crossover frequency of approximately 0.25 Hz, with a small phase margin of less than 10 degrees. With this control performance, larges transient response time and high overshoot and ringing will be expected in $v_{do}(t)$ for a step change in EV-parameters.

Let's now include a compensator system eq. (20) into eq. (17) to achieve a sufficiently large crossover frequency and phase margin. For example, let's achieve a crossover frequency of a least one tenth of the switching frequency and a phase margin higher than 60 degrees. The zero and pole frequencies in eq. (20) can be obtained using eq. (21) and (22). The constant gain is obtained using eq. (23), where T_{u0} is the low-frequency magnitude of the uncompensated loop gain; f_L in eq. (20) can be chosen arbitrarily, for example f_L= one-tenth of the crossover frequency or 35 Hz; and $f_0 = 1/2\pi(\sqrt{L_{eq}C})$.

$$f_z = (0.35kHz)\sqrt{\frac{1 - \sin(60°)}{1 + \sin(60°)}} \approx 0.1kHz \quad (21)$$

$$f_p = (0.35kHz)\sqrt{\frac{1 + \sin(60°)}{1 - \sin(60°)}} = 1.3kHz \quad (22)$$

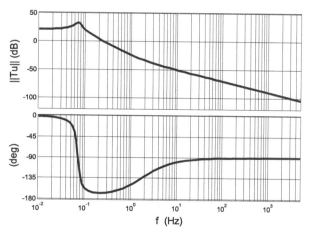

Figure 4: Characteristic of the uncompensated loop gain with $G_c(s) = 1$.

$$K_c = \left(\frac{f_c}{f_0}\right)^2 \frac{1}{T_{u0}} \sqrt{\frac{f_z}{f_{p1,2}}} \quad (23)$$

The characteristic of the compensated loop gain is sketched in Figure 5. It can be seen that the phase margin is larger than 60 degrees over the frequency range from 100 Hz to high-frequency. The crossover frequency is moved from 0.25 Hz (in Figure 4) to 0.35 kHz or 350 Hz in Figure 5, thus the transient response time is shortened. Hence variations in i_a should have little impact on the phase margin. Also, the low-frequency voltage loop-gain magnitude has increased, from about 20 dB in Figure 4 to 100 dB in Figure 5, thus reducing effectively the magnitude of $Z_a(s)$ and $G_{vb}(s)$ in eq. (15) and (16).

Control system performance evaluation

Step changes or disturbances of magnitude $\|\hat{v}_{do}\|$ and $\|\hat{i}_a\|$ are respectively applied to the proposed scheme in Figure 3. The characteristics shown in Figures 6 and 7

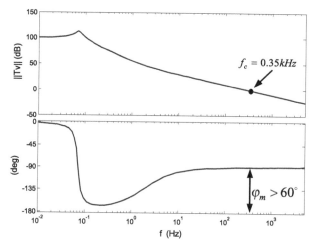

Figure 5: Characteristic of the compensated loop gain $T_v(s)$ with $G_c(s) =$eq. (20).

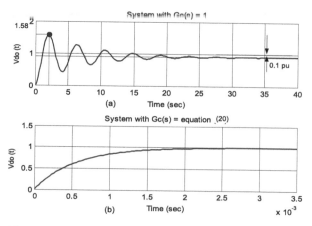

Figure 6: Characteristics of the voltage signal at the input of the DC motor when small variations of parameters (current, voltage, duty ratio) occur in the EV-power grid, (a) system with no compensator showing high overshoot, ringing, and steady-state error (b) system with a compensator showing no overshoot, ringing, and zero steady-state error.

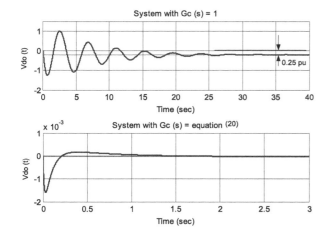

Figure 7: Characteristics of the voltage signal at the input of the DC motor when a step change occurs in BSS output voltage: (a) system with no compensator showing high overshoot, ringing, and steady-state error; (b) system with a compensator showing no overshoot, ringing, and zero steady-state error.

are obtained, respectively, for the system with no compensator ($G_c(s) = 1$) and for the system with a compensator ($G_c(s) = $ eq. (20)).

It can be seen from Figure 6(b) and 7(b) that the proposed controller scheme, which includes the compensator or controller $G_c(s) = $ eq. (20) in the closed-loop, maintains well the voltage variations at the input of the DC motor with zero steady-state error, no overshoot and ringing, in spite of step changes in EV-power grid parameters.

Conclusion

The evaluated EV-power grid scheme, consisting of: 'a BSS, an isolated flyback DC-DC converter with a controller, and a DC motor' performs well in maintaining the input voltage of the DC motor within acceptable limits of variation in spite of the variations of parameters in the EV-circuits and the effect of BSS charging/discharging process. No overshoot and ringing are observed in the DC voltage characteristics; also, the steady-state error is kept to about zero. Further analyses can be conducted using PSIm or PSCAD to investigate other performances such as the fault ride through capability.

ORCID

Mammo Muchie ⓘ http://orcid.org/0000-0003-4831-3113

References

Albiol-Tendillo, L., E. Vidal-Idiarte, J. Maixé-Altés, J. M. Bosque-Moncusí, and H. Valderrama-Blavi. 2012. "*Design and Control of a Bidirectional DC/DC Converter for an Electric Vehicle.*" Paper presented at the 15th international power electronics and motion control conference (EPE/PEMC), Novi Sad, Serbia, 4–6 Sept. 2012.

Anwar, S., W. Zhang, F. Wang, and D. J. Costinett. 2016. "*Integrated DC-DC Converter Design for Electric Vehicle Power Trains.*" Paper presented at the IEEE applied power electronics conference and exposition (APEC), Long Beach, USA, 20–24 March 2016.

Biradar, S. K., R. A. Patil, and M. Ullegaddi. 1998. "*Energy Storage System in Electric Vehicle.*" Paper presented at the power quality conference, Hyderabad, India, 18 June 1998.

Erickson, R. W. 2001. *Fundamentals of Power Electronics*. New York: Springer Publishers.

Huang, Q., J. Li, and Y. Chen. 2010. "Control of Electric Vehicle." In *Urban Transport and Hybrid Vehicles*, edited by Seref Soylu, Chap. 9. Sciyo, Croatia: Intech Open Science Publishers.

Kumar, A., and P. Gaur. 2014. "*Bidirectional DC/DC Converter for Hybrid Electric Vehicle.*" Paper presented at the international conference on advances in computing, communications and informatics (ICACCI), 24–27 Sept. 2014.

Lovatt, H. C., V. S. Ramsden, and B. C. Mecrow. 1998. "Design of an in-Wheel Motor for a Solar-Powered Electric Vehicle." *IEE Proceedings - Electric Power Applications* 145 (5): 402–408.

Mohan, N., T. M. Undeland, and W. P. Robbins. 2003. *Power Electronics: Converters, Applications, and Design*. New York: John Wiley & Sons, Inc.

Momoh, O. D., and M. L. Omoigui. 2009. "*An Overview of Hybrid Electric Vehicle Technology.*" Paper presented at the 5th IEEE vehicle power and propulsion conference, Dearborn, Michigan, September 7–11, 2009.

Musasa, K., M. N. Gitau, and R. C. Bansal. 2015a. "Performance Analysis of Power Converter Based Active Rectifier for an Offshore Wind Park." *Electric Power Components and Systems* 43 (8–10): 1089–1099.

Musasa, K., M. N. Gitau, and R. C. Bansal. 2015b. "Dynamic Analysis of DC–DC Converter Internal to an Offshore Wind Farm." *IET Renewable Power Generation* 9 (6): 542–548.

Pany, P., R. K. Singh, and R. K. Tripathi. 2011. "Bidirectional DC-DC Converter fed Drive for Electric Vehicle System." *International Journal of Engineering, Science and Technology* 3 (3): 101–110.

Sankara, A. B., and R. Seyezhai. 2016. "Simulation and Implementation of Solar Powered Electric Vehicle." *Circuits and Systems* 7: 643–661.

Santos, R., F. Pais, C. Ferreira, H. Ribeiro, and P. Matos. 2007. "Electric Vehicle – Design and Implementation Strategies for the Power Train." *Renewable Energy & Power Quality Journal* 1 (5): 552–558.

Veilleux, E., and P. W. Lehn. 2014. "Interconnection of Direct-Drive Wind Turbines Using a Series-Connected DC Grid." *IEEE Transactions on Sustainable Energy* 5 (1): 139–147.

Young, K., C. Wang, L. Y. Wang, and K. Strunz. 2013. "Electric Vehicle Battery Technologies." In *Electric Vehicle Integration Into Modern Power Networks*, edited by R. Garcia-Valle, and J. A. Peças Lopes, 15–56. New York: Springer Publishers.

Temporal analysis of electricity consumption for prepaid metered low- and high-income households in Soweto, South Africa

Njabulo Kambule, Kowiyou Yessoufou, Nnamdi Nwulu and Charles Mbohwa

This study explores the temporal trend in electricity consumption since the introduction of prepaid metres in low-income households of Soweto and compares the findings with high-income households. Monthly electricity consumption data (over 96 months: 2007–2014) for 4427 households in Soweto, for both low- and high-income households, was collected from Eskom. Using a simple linear model to analyse consumption trends in low-income households, we ascertained that electricity consumption has decreased by 48% since the inception of prepaid metres. Nonetheless, it is noted that 60% of household incomes are spent on electricity bills, which is way above the threshold set for energy poverty. Comparatively, high-income households consume less electricity than low-income households do. Overall, the prepaid metre programme is producing expected results for Eskom but remains a challenge for low-income households, which are still entrenched in energy poverty. We call for an energy policy that is tailored for each income groups and the formulation of laws and policies to protect the energy vulnerable households.

Introduction

Household electricity debt continues to plague South Africa's electricity supplying utility – Eskom – as only 16% of households pay for the electricity service Eskom provides (Timeslive 2015). The consequence of non-payment on public utilities is enormous. First, it is an important constraint to the provision of electricity services (Szabo and Ujhelyi 2014). Second, the shortage in revenue that comes with non-payment ultimately results in maintenance backlogs, system deterioration, inability to purchase fuel to operate the generating units, and the deterioration of the economy (World Bank 1999). Third, studies have found that there is a correlation between non-payment and expenditure ratios: households with increasing electricity consumption (as a percentage of total household expenditure) are more likely to not pay regularly for their consumption (World Bank 1999; Lampietti, Banerjee, and Branczik 2007).

Different countries experience the problem of non-payment differently and deal with it in different ways. In 2003, the South African government persuaded Eskom to erase electricity household debt of R1.4 billion (US $1.2 million), under the Free Basic Electricity (FBE) policy (Styan 2015). Then, the government took some measures to avoid further debt in the future. These measures included the introduction of prepaid meters and a free monthly allocation of electricity incentive (50 kWh) for all indigent households that agreed to have prepaid meters installed (DME 2003). As opposed to the expected outcome of these policies (e.g. Free Basic Electricity, free 50 kWh electricity consumption), household debt increased to R13.6 billion (US$1.14 billion), of which Soweto township alone owed R8.6 billion (US $7.2 million), 60% of the total amount due to massive electricity non-payment (Styan 2015). As a result,

Eskom has resorted to intensifying its efforts to deploy prepaid household meters in the township.

In 2007, the first official prepaid electricity-meter pilot project was undertaken in a small region of Soweto known as *Chiawelo*. To date, more than 45% of low-income households in Soweto are prepaid metered; the target is that all households be connected to this model of payment by 2020 (City Press 2016). While households have protested and rejected the prepaid meter technology, Eskom has persistently echoed that the technology stands to benefit rather than harm households (Ruiters 2007; Makonese, Kimemia, and Annergarn 2012; Chinomona and Sandada 2014; City Press 2015; Jack and Smith 2015, 2016; Press Reader 2015; Timeslive 2015; SABC 2016; IOL 2017) and this view is supported by several studies across various geographic regions where significant decreases in electricity consumption were reported (Darby 2006; Fischer 2008; Faruqui, Sergici, and Sharif 2010; Gans, Alberini, and Longo 2013; Qui and Xing 2015). For example, Faruqui, Sergici, and Sharif (2010) found that in North America the prepaid-metre technology reduced electricity consumption by 7% whereas in Northern Ireland, Gans, Alberini, and Longo (2013) reported a decrease of 11–17%. Similarly, Martin (2014) reported that households using prepaid electricity metres in Kentucky, USA, have reduced their electricity consumption by 11%. A study conducted in Canada noted that 25% of the sampled households utilized 20% less electricity with prepaid-metre technology (Casarin and Nicollier 2009). This decrease is the result of direct feedback on electricity consumption provided by prepaid meters, thus enabling consumers to monitor their electricity consumption.

Although Qui and Xing (2015) confirmed the reduction pattern of electricity consumption in Arizona,

USA, due to prepaid metres, they also called for caution in generalizing this trend as they pointed out that socio-economic factors matter in electricity consumption. Specifically, they indicated that low-income households tend to experience more electricity reductions than high-income households, suggesting a context-dependent effect of prepaid meters on electricity consumption. Based on Qui and Xing's (2015) finding, we therefore hypothesized that low-income households in Soweto would also consume more electricity than high-income households due to the differences in the types of appliances used in both income groups. This is in line with several other studies that questioned whether prepaid meters are truly beneficial to low-income households (Colton 2001; Ruiters 2007; van Heusden 2010; O'Sullivan, Howden-Chapman, and Fougere 2011; Hittinger et al. 2012; Makonese, Kimemia, and Annergarn 2012; Malama et al. 2014). These studies have alluded to the fact that prepaid electricity meters have the potential of entrenching energy poverty, especially among energy-vulnerable households. Energy vulnerability precedes energy poverty. At the vulnerable phase, there is a set of prevailing household conditions or factors that may lead to poverty. An acknowledgement of this phase helps to identify groups of people that may be at risk of being energy-poor in the near future. In expenditure terms, these are households that tend to spend more than 10% of their income to meet their energy-related needs (O'Sullivan, Howden-Chapman, and Fougere 2011; Bouzarovski and Petrova 2015; Ismail and Khembo 2015). Poor household energy efficiency, increasing electricity costs, and overcrowding are identified as other key causal factors in energy poverty (The Guardian 2016).

For example, when, from 1990 to 2008, the price of electricity in New Zealand markedly increased by 71%, this increase led to energy poverty in low-income areas but not in high-income households (O'Sullivan, Howden-Chapman, and Fougere 2011, 2015). The differences in electricity consumption in low- versus high-income households are due to the differences in the nature of dwellings and appliances used (Genjo et al. 2005; Tso and Yau 2007; Druckman and Jackson 2008; Wiesmann et al. 2011; DoE 2012; Bedir, Hasselaar, and Itard 2013; Jones, Fuertes, and Lomas 2015). In such a context, it is not an effective measure to design a 'one-size-fit-all policy' for electricity consumption across all households without taking into consideration socio-economic differences.

The gap the present study fills, and its contribution are as follows. In the context of Soweto, we do not know the consumption patterns over time in low- versus high-income households and how consumption in high-income households compares to consumption in low-income households. The present study provides answers to this gap in the information we have, and, in so doing, contributes knowledge that can inform the prepaid meter programme in Soweto. Specifically, we aim to understand the influence of prepaid meters on electricity consumption in this South Africa's township. Our objectives are three-fold: i) to identify the trend of electricity consumption since the introduction of prepaid metre in 2007; ii) to identify the proportion of change in consumption since 2007; and iii) to compare electricity consumption in low- versus high-income households in Soweto township.

Material and method
Study area
The present study was conducted in Soweto, the largest (200.3 km^2) township in South Africa with a population of about 1,271,628 inhabitants (Frith 2017). Within the township, two areas were targeted, namely, Chiawelo and Diepkloof Extension.

Chiawelo is a largely low-income household area of Soweto, established in 1956. Its size is about 1.10 km^2, and was developed to provide cheap accommodation for black workers (specifically Tsonga- and Venda-speaking South Africans) during the apartheid era. Approximately 3841 households are found in Chiawelo (Frith 2017). The household structures have 3–4 small rooms (with each room size of ~ 32 m^2). A significant majority of these households generate money by renting backyard dwellings (e.g. shacks they have constructed). Eskom installed prepaid meters in the area in 2007. The area is still considered one of the socio-economically disadvantaged areas in Soweto.

Diepkloof Extension (DE) is a segment region of Diepkloof sub-township with a surface area of about 1.42 km^2 (Frith 2017). It has about 1564 households. The area developed in the early 1980s and 1990s (Alexander et al. 2013). It was built for middle- to upper-class blacks (wealthier blacks, who were largely professionals employed by the state, but also privately employed professionals) (Marx and Rubin 2008; Alexander et al. 2013). The area is therefore referred to as the 'Rich Man's Acre'; the house structures are bigger and intended as a more exclusive area. Most households in DE are modern and constantly being renovated towards being energy efficient. They are noticeably distinct from houses in the old townships of Soweto, with structures that are more permanent, built with expensive building materials (brick and tiled roofs as opposed to corrugated iron walls and roofs). The houses' physical structure size is estimated to be between 200 and 300 m^2. The area received prepaid meters in 2013. Overall, this is a socio-economically well-off region of Soweto.

Data collection
The electricity consumption data in Chiawelo and DE was acquired from Eskom. The agreement between the utility and the researchers is that the anonymity of each household's data (electricity consumption and cost) is preserved. We maintained this anonymity by not sharing the collected raw data for any household. Prepaid meters were not installed in both Chiawelo and DE at the same time. While the prepaid metre programme was introduced to low-income households in 2007, the programme was introduced to high-income households only in 2016. As a result, data on consumption were not available for similar period. Data on electricity consumption and cost for Chiawelo (low-income households) were collected monthly over 96 months i.e. eight years (2007 to 2014) for 3 841 households whereas data for DE (high-income

households) were collected for 1 564 total households only from June 2016 to February 2017 for the reason indicated above (i.e. prepaid meters were introduced to high-income households in 2016 and data for this income group are available for only 2016 to 2017).

Data analysis

All analyses were done in R (R-Development Core Team 2015). Firstly, we analyzed the trend of electricity consumption over eight years (2007–2014) in low-income households. This analysis was done using a simple linear regression. Secondly, we assessed by how much the electricity consumption had changed in low-income households over the study period (2007–2014) since the introduction of prepaid meters to this income group. This change was calculated as:

$C_{consumption} = ((C_{2014} - C_{2007})/C_{2007})$; C_{2014} and C_{2007} are the total electricity consumption in 2014 and 2007, respectively.

We then analyzed the trend in monthly electricity consumption over the same period using the analysis of covariance (ANCOVA) with year and month as co-variates. To further understand the monthly patterns of electricity consumption, we ran a one-way ANOVA using consumption as a response variable and month as a predictive variable.

Finally, we compared the consumption patterns between low- and high-income households. Because we have monthly consumption data for only a limited period of time (nine months: June 2016 to February 2017) in high-income households while we have monthly consumption for 96 months in low-income households, we tested the differences in consumption in both income groups by comparing the average monthly consumption in low-income households versus the average monthly consumption in randomly drawn years from 2007 to 2014 in low-income households. This comparison was done as follows. We selected randomly 100 times a year between 2007 and 2014, and calculated the average monthly consumption for each randomly selected year. Then, we calculated the actual monthly average consumption of high-income households over the period of June 2016 to February 2017 and compared this actual average to the average monthly consumption in low-income households of the randomly selected years. The significance of the difference between actual and random consumption was assessed using the 95% confidence interval (CI). For this particular analysis, the consumption data were $\log_{10}$-transformed to meet the normal error distribution.

Results

Our results indicate that electricity consumption in low-income households decreased significantly ($P < 0.001$) over the study period (2007–2014), following the trend $y = -63.78 x + 129071.94$ ($R^2 = 25.13\%$) (Figure 1). We also found that, since the introduction of prepaid electricity meters, the consumption levels decreased by a monthly average of 48% in low-income households over the study period. The decrease reached its lowest level in 2010 (610–930 kWh) (Figure 2). However, this overall decreasing trend hides some specificities in some months where the consumptions increased significantly in comparison to the average monthly consumption. These months include May ($\beta = 252.94 \pm 122.62$; $P = 0.04$), June ($\beta = 436.82 \pm 118.72$; $P = 0.0004$), July ($\beta = 453.73 \pm 118.72$; $P = 0.0002$) and August ($\beta = 310.69 \pm 118.72$; $P = 0.01$) (Figure 3). Finally, we found that energy consumption in low-income households was significantly higher than consumption in high-income household [(mean consumption in high-income household (log) $= 6.28$; CI $= 6.65$–6.76)] (Figure 4). Our model [$y = -63.78 x + 129071.94$ ($R^2 = 25.13\%$)] shows that the low-income household electricity consumption levels equalled high-income household levels in the year 2015.

Discussion

As a means to reduce the challenge of household electricity non-payment in Soweto, Eskom effected the process of prepaid electricity meter deployment in 2007. This study evaluates the role of the technology, particularly regarding electricity consumption, in low- and high-income households in the township. We firstly assessed the trend in prepaid electricity consumption in low-income households in Soweto. Using a simple linear model, we found that the rate of consumption has decreased by 48% between 2007 and 2014. We acknowledge that the strength of our model is weak (25%), suggesting that 75% of variation in electricity consumption remains unexplained by the linear model. However, our finding is broadly consistent with the general trend reported in several studies across various geographic regions (Canada: 20% decrease, Casarin and Nicollier 2009; North America: 7% decrease, Faruqui, Sergici, and Sharif 2010; Northern Ireland: 11–17% decrease; Gans, Alberini, and Longo 2013; Kentucky: 11% decrease; Martin 2014; Arizona: 12% decrease, Qui and Xing 2015). It is also important to highlight that the decrease reported in our study is much higher (48%) than any other decrease reported elsewhere in relation to the installation of prepaid meters. This is potentially due to the fact that we focused only on electricity consumption in low-income households, whereas other studies analyzed a combined dataset from both low- and high-income households (Qui and Xing 2015).

Our results also reflect that the overall declining trend camouflages some energy consumption specificities in some months. For example, despite the overall decreasing trend, consumption increased significantly in comparison to the average monthly consumption from May to August, with the lowest consumption observed in June 2010. The significant increased consumption in May–August is linked to the winter period in South Africa when the response to the cold weather requires an increase in energy demand to warm houses. Households therefore spend more and become more energy vulnerable during the winter season (Lampietti, Banerjee, and Branczik 2007). The sharp decline in electricity consumption in June 2010 in Soweto's low-income households could be linked to the Soccer World Cup. Soccer in Soweto is almost a religion, and the world cup was an opportunity for people to experience the once-in-a-lifetime opportunity

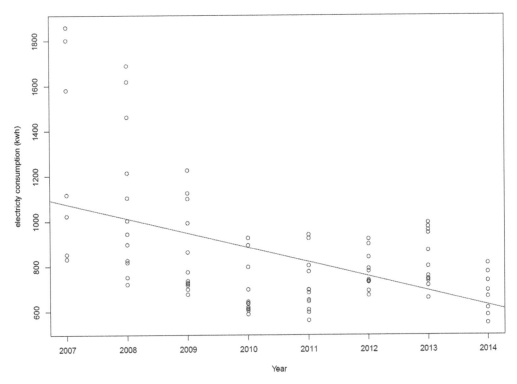

Figure 1: Trend of yearly electricity consumption in low-income households since the introduction of prepaid meters in 2007 (2007–2014).

that the world cup provided. As such, people were mostly outdoors in different stadiums watching the games, or people convened at one single household to watch the games, and consequently the consumption of electricity dropped significantly explaining the lowest June 2010 energy consumption that we observed in our study. This is an illustration of how social events such as the soccer world cup may contribute to energy efficiency in low-income households. However, it is also important to find out if this is true at a national level.

The overall decreasing trend should not mask the socio-economic implications for low-income households. Understanding these socio-economic implications is very important in a policymaking process that takes income level into consideration. For example, each household in Chiawelo – the low-income region in this study – consumed an average

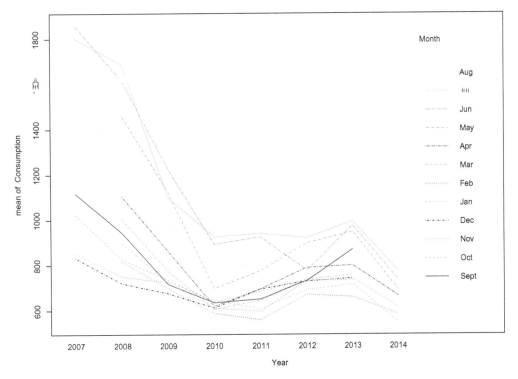

Figure 2: Trend of yearly electricity consumption in low-income households highlighting the monthly patterns.

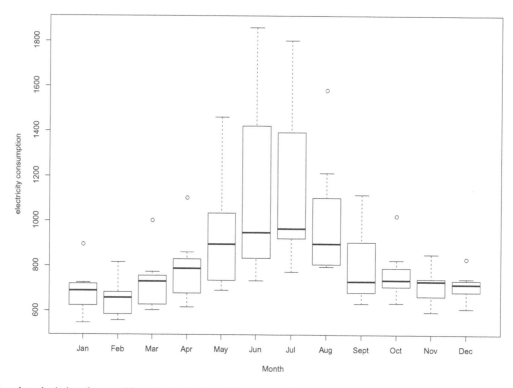

Figure 3: Boxplots depicting the monthly variation in electricity consumption.

of 667.6 kWh in 2014. Assuming the 2014 electricity tariff of R0.98 per kWh, households therefore monthly spent R654.2. By applying the upper-bound poverty line (UBPL) (using the 2015 prices), low-income households have a monthly income of approximately R992 per month (StatsSA 2017), meaning that households spent about 66% of their monthly income just to cover their electricity consumption. This renders households vulnerable to energy poverty, particularly in the context of recent and predicted future increases in electricity tariffs. Consequently, prepaid meters may lead to decreased electricity consumption, but it does not solve the problem of energy poverty in low-income households. This is in support of the previous views that prepaid meters may entrench socio-economic marginalization and electricity inequality (Colton 2001; Ruiters 2007; van Heusden 2010), given that spending up to 46% of households income (in winter) only on electricity consumption would have severe consequence on many other sectors of the household lives including education, food, health, clothing, transport, etc. This is in contrast to an early claim that prepaid meters improve social welfare (Casarin and Nicollier 2009).

To alleviate the weight of electricity consumption, the incentive measures put in place by the government include free monthly consumption of 50 kWh. This is clearly not enough in light of the amount spent to cover electricity bills despite the remarkable decrease in consumption. The current energy or electricity policy landscape for households is poor and offers only limited energy security to the poor. It does not offer energy vulnerable or impoverished households protection. The FBE policy is currently the main policy that provides an incentive for poor households. We strongly argue that it lacks relevance and needs to be evaluated and updated according to the socio-economic realities faced by indigent households.

Similar evidence of energy poverty has also been reported in many other countries. In Zambia, for example, Malama et al. (2014) reported that low-income households suffer from prepaid metre disconnections because of high and unaffordable prices (Malama et al. 2014). In response, low-income households shift from using electricity to alternative energy sources such as wood or coal-burning stoves, paraffin fuelled heaters, gas heaters, hot water bottles and usage of bed-blankets. This is an important finding for policymakers because it carries indicators closely associated with fuel or energy poverty – that is the inability of households to acquire adequate household electricity for safe and healthy indoor temperatures (O'Sullivan et al. 2013). Furthermore, according to Ismail and Khembo (2015), the energy poverty expenditure line is estimated to be 10–15% of income. Low-income households were found to spend about 66% of their income on electricity. This is an apparent indicator of socio-economic marginalization of poor households. Higher expenditure on electricity among low-income households means they become more energy vulnerable (World Bank 1999; Lampietti, Banerjee, and Branczik 2007). Again, with increasing unemployment and electricity tariffs (by 400% in the past decade) in Soweto, the ill effect of prepaid meters on energy poverty needs to be thoroughly studied. There is a need for policymakers to re-assess and monitor the prepaid meter programme, and, as Colton (2001) advised, establish mechanisms (e.g. laws) to protect the fuel or energy vulnerable and impoverished households.

How does electricity consumption in low-income households compare to the consumption in high-income households? Our analysis showed that low-income households consume significantly more electricity than high-income households despite the trend towards a decreased

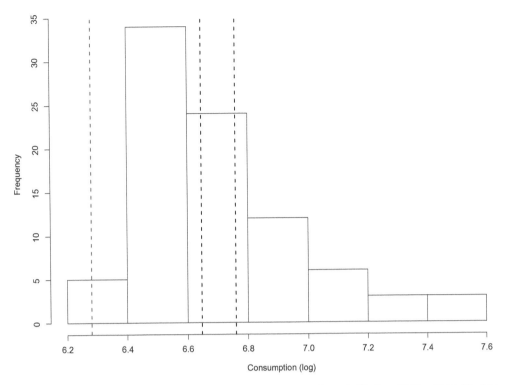

Figure 4: Comparison of electricity consumption between low- and high-income households from 2007 to 2014. The histogram indicates the average distribution of monthly electricity consumption in randomly selected years in low-income households between 2007 and 2014. The bold dotted black lines indicate the confidence interval for the random consumption. The bold dotted line indicates the actual monthly average consumption of high-income households. The dotted black and vertical lines mean: 1) Black dotted line – confidence interval for consumption in low-income household. 2) Vertical dotted line – average monthly consumption of high-income households.

consumption in low-income households over several years. This may largely be attributable to factors unique to low-income households, for instance their continued dependence on old energy inefficient appliances and provision of space to rent backyard rooms or shacks on their premises (Makonese, Kimemia, and Annergarn 2012) – as compared to energy efficient buildings and appliances used by high-income households. A study conducted by Parker (2003) reported that because of *inter alia* less efficient appliances, older homes consumed greater electrical energy for space heating and cooling. Furthermore, low-income households are also characterized by backyard dwellings that are rented out, thus contributing to additional electricity consumption.

In contrast, high-income households comprise in general only two employed persons per household (StatsSA 2011). This household category receives an annual income of more than R307 201 (about R26 000 per month). This household is characterized by a dwelling solely dependent on electricity and gas as the main sources of energy for cooking, lighting and heating. Electricity appliances found in households in this income group include a radio, television, computer, refrigerator and cell phone. This household category is considered broadly energy efficient and can afford energy efficient electrical appliances, thus justifying the lower electricity consumption in comparison to low-income households. Several other factors not explicitly explored in the present study may also account for the differences observed. These include household expenditure patterns,

education level, household and dwelling size, location of the household, all factors that have been recently identified as determinants of energy poverty in South Africa (Ismail and Khembo 2015). Based on our simple linear model, we identified that low-income households may have reached the same consumption level as high-income households since 2015. Unfortunately, we do not yet have a consumption dataset beyond 2014 to confirm this prediction, which also precludes us from verifying whether the general decreasing trend since 2007 has been maintained beyond 2014.

Conclusion

South Africa's Energy White Paper clearly indicated that 'energy security for low-income households can help reduce poverty, increase livelihoods and improve living standards' (DME 1998). Our trend analysis reveals that low-income households are consuming lesser electricity over time owing to prepaid meters, and this is positive development in support of the prepaid meter policy established since 2007. However, as more than 60% of the income in indigent households is spent on electricity consumption, way above the energy poverty expenditure threshold, estimated to be 10–15% (Ismail and Khembo 2015). We recommend to policymakers to review and monitor the prepaid metre programme and to formulate tools (e.g. laws) to protect such energy vulnerable and impoverished households. Furthermore, we conclude that despite this decrease in energy consumption, low-income households continue to consume more electricity than high-income households. We therefore propose that in the midst of the current electricity crisis in South Africa, there is an urgent need for the government to subsidize the installation of renewable energy technologies in low-income households. We also recommend that since the FBE policy is currently the main document providing an incentive to poor households, this policy needs to be evaluated and updated according to the socio-economic and energy poverty realities facing the majority of low-income households in Soweto.

Acknowledgements

We would like to thank Eskom officials, namely Nathi Motsoane and Thandazile Mazibuko, who willingly provided all the data used in this study and for their insights shared regarding some of the findings highlighted in this study.

Disclosure statement

No potential conflict of interest was reported by the authors.

References

Alexander, P., C. Ceruti, K. Motseke, M. Phadi, and K. Wale. 2013. *Class in Soweto*. Durban: University of KwaZulu-Natal Press.

Bedir, M., E. Hasselaar, and L. Itard. 2013. "Determinants of Electricity Consumption in Dutch Dwellings." *Energy and Buildings* 58: 194–207.

Bouzarovski, S., and S. Petrova. 2015. "A Global Perspective on Domestic Energy Deprivation: Overcoming the Energy Poverty–Fuel Poverty Binary." *Energy Research & Social Science* 10: 31–40.

Casarin, A. A., and L. Nicollier. 2009. Prepaid Meters in Electricity. A Cost-Benefit Analysis. Working Paper Series IAE.

Chinomona, R., and M. Sandada. 2014. "Customers' Perceptions on Eskom's Pre-Paid Billing System and the Effects on Their Satisfaction and Trust." *Mediterranean Journal of Social Sciences* 9: 119–126.

City Press. 2015. Soweto Residents Want Flat Electricity Rate But Use Energy Hungry Appliances. Accessed December 13, 2017. http://city-press.news24.com/News/Soweto-residents-want-flat-electricity-rate-but-use-energy-hungry-appliances-20150510.

City Press. 2016. Eskom Holds Thumbs Soweto Will Accept its New Smart Meters. Accessed September 20, 2017. http://city-press.news24.com/News/eskom-holds-thumbs-soweto-will-accept-its-new-smart-meters-20160430.

Colton, D. R. 2001. "Prepayment Utility Meters, Affordable Home Energy, and the Low Income Utility Consumer." *Journal of Affordable Housing and Development Law* 3: 285–305.

Darby, S. 2006. The Effectiveness of Feedback on Energy Consumption: A Review for Defra of the Literature on Metering, Billing and Direct Displays. Page: 1–21.

DME (Department of Minerals and Energy). 1998. White Paper on the Energy Policy of the Republic of South Africa. Available Online: http://www.energy.gov.za/files/policies/whitepaper_energypolicy_1998.pdf.

DME (Department of Minerals and Energy). 2003. Electricity Basic Services Support Tariff (Free Basic Electricity) Policy. Accessed March 12, 2018. http://www.energy.gov.za/files/policies/Free20Basic20Electricity20Policy202003.pdf.

DoE (Department of Energy). 2012. A Survey of Energyrelated Behaviour and Perceptions in South Africa: The Residential Sector. Accessed at http://www.energy.gov.za/files/media/Pub/Survey%20of%20Energy%20related%20behaviour%20and%20perception%20in%20SA%20-%20Residential%20Sector%20-%202012.pdf

Druckman, A., and T. Jackson. 2008. "Household Energy Consumption in the UK: A Highly Geographically and Socio-Economically Disaggregated Model." *Energy Policy* 36: 3177–3192.

Faruqui, A., S. Sergici, and A. Sharif. 2010. "The Impact of Informational Feedback on Energy Consumption—A Survey of the Experimental Evidence." *Energy* 35: 1598–1608.

Fischer, C. 2008. "Feedback on Household Electricity Consumption: A Tool for Saving Energy?" *Energy Efficiency* 1: 79–104.

Frith, A. 2017. Population Statistics. Accessed March 12, 2018. https://census2011.adrianfrith.com/place/798026.

Gans, W., A. Alberini, and A. Longo. 2013. "Smart Meter Devices and the Effect of Feedback on Residential Electricity Consumption: Evidence from a Natural Experiment in Northern Ireland." *Energy Economics* 36: 729–743.

Genjo, K., S.-i. Tanabe, S.-i. Matsumoto, K.-i. Hasegawa, and H. Yoshino. 2005. "Relationship Between Possession of Electric Appliances and Electricity for Lighting and Others in Japanese Households." *Energy and Buildings* 37: 259–272.

Hittinger, E., A. Kimberley, I. Mullins, and L. Azevedo. 2012. "Electricity Consumption and Energy Savings Potential of Video Game Consoles in the United States." *Energy Efficiency*, doi:10.1007/s12053-012-9152-z.

IOL. 2017. Chiawelo Residents Stick to Demands. Accessed September 19, 2017. https://www.iol.co.za/news/south-africa/gauteng/chiawelo-residents-stick-to-demands-1094881.

Ismail, Z., and P. Khembo. 2015. "Determinants of Energy Poverty in South Africa." *Journal of Energy in Southern Africa* 3: 66–78.

Jack, B. K., and G. Smith. 2015. Pay as You Go: Pre-paid Metering and Electricity Expenditure in South Africa. Accessed November 15, 2016. https://sites.tufts.edu/kjack/files/2015/08/Jack_manuscript7.pdf.

Jack, B. K., and G. Smith. 2016. Charging Ahead: Prepaid Electctricity Metering in South Africa. NBER Working Paper Series.

Jones, R. V., A. Fuertes, and K. J. Lomas. 2015. "The Socio-economic, Dwelling and Appliance Related Factors Affecting Electricity Consumption in Domestic Buildings." *Renewable and Sustainable Energy Reviews* 43: 901–917.

Lampietti, J. A., S. G. Banerjee, and A. Branczik. 2007. People and Power Electricity Sector Reforms and the Poor in Europe and Central Asia. https://openknowledge.worldbank.org/handle/10986/7175.

Makonese, T., D. K. Kimemia, and H. J. Annergarn. 2012. Assessment of Free Basic Electricity and Use of Pre-paid Meters in South Africa. Accessed October 13, 2016. http://conferences.ufs.ac.za/dl/Userfiles/Documents/00000/577_eng.pdf.

Malama, A., P. Mudenda, A. Ng'ombe, L. Makashini, and H. Abanda. 2014. "The Effects of the Introduction of Prepayment Meters on the Energy Usage Behaviour of Different Housing Consumer Groups in Kitwe, Zambia." *AIMS Energy* 2: 237–259.

Martin, M. W. 2014. Pay As You Go Electricity: The Impact of Prepay Programmes on Electricity Consumption. Theses and Dissertations – Agricultural Economics. p. 29. Accessed October 25, 2016. http://uknowledge.uky.edu/agecon_etds/29.

Marx, C., and M. Rubin. 2008. 'Divisible Spaces': Land Biographies in Diepkloof, Thokoza and Doornfontein, Gauteng. A final draft report prepared for Urban LandMark.

O'Sullivan, K. C., P. L. Howden-Chapman, and G. Fougere. 2011. "Making the Connection: The Relationship Between Fuel Poverty, Electricity Disconnection, and Prepayment Metering." *Energy Policy* 39: 733–741.

O'Sullivan, K. C., P. L. Howden-Chapman, and G. M. Fougere. 2015. "Fuel Poverty, Policy, and Equity in New Zealand:

The Promise of Prepayment Metering." *Energy Research & Social Science* 7: 99–107.

O'Sullivan, K. C., P. L. Howden-Chapman, G. M. Fougere, S. Hales, and J. Stanley. 2013. "Empowered? Examining Self-disconnection in a Postal Survey of Electricity Prepayment Meter Consumers in New Zealand." *Energy Policy* 52: 277–287.

Parker, D. S. 2003. "Research Highlights from a Large Scale Residential Monitoring Study in a Hot Climate." *Energy and Buildings* 35: 863–876.

Press Reader. 2015. Why Sowetan's Won't Pay. Accessed September 19, 2017. https://www.pressreader.com/south-africa/sowetan/20150518/281599534080349.

Qui, L., and B. Xing. 2015. Pre-paid Electricity Plan and Electricity Consumption Behaviour. Accessed August, 18, 2017. http://cmepr.gmu.edu/wp-content/uploads/2014/01/Qiu_pre-paid-pricing_0202.pdf.

R Development Core Team. 2015. *R: A Language and Environment for Statistical Computing*. Vienna: R Foundation for Statistical Computing.

Ruiters, G. 2007. Free Basic Electricity in South Africa: A Strategy for Helping People or Containing the Poor?

SABC (South African Broadcasting Corporation). 2016. 40 000 Smart-Prepaid Metering Systems Installed in Soweto. Accessed February 20, 2017. http://www.sabc.co.za/news/a/778df0804c949ca0b355bfa10755317a/40-000-Smart-Prepaid-metering-systems-installed-in-Soweto.

StatsSA (Statistics South Africa). 2011. Income Dynamics and Poverty Status of Households in South Africa. Accessed September 21, 2017. http://www.statssa.gov.za/publications/Report-03-10-10/Report-03-10-102014.pdf.

StatsSA (Statistics South Africa). 2017. Poverty on the Rise in South Africa. Accessed March 13, 2018. http://www.statssa.gov.za/?p=10334.

Styan, J. B. 2015. *Blackout: The Eskom Crisis*. Johannesburg & Cape Town: Jonathan Ball Publishers.

Szabo, A., and G. Ujhelyi. 2014. Can information Reduce Non-payment for Public Utilities? Experimental Evidence from South Africa. ftp://ftp.repec.org/opt/ReDIF/RePEc/hou/wpaper/2014-114-31.pdf

The Guardian. 2016. The Three Housing Problems that Most Affect Your Health. https://www.theguardian.com/society-professionals/2014/aug/08/housing-problems-affect-health.

Timeslive. 2015. Soweto Shrugs off R4 Billion Eskom Bill. Accessed February 14, 2017. http://www.timeslive.co.za/local/2015/02/03/Soweto-shrugs-off-R4-billion-Eskom-bill.

Tso, G. K. F., and K. K. W. Yau. 2007. "Predicting Electricity Energy Consumption: A Comparison of Regression Analysis, Decision Tree and Neural Networks." *Energy* 32: 1761–1768.

van Heusden, P. 2010. Discipline and the New 'Logic of Delivery': Prepaid Electricity in South Africa and Beyond.

Wiesmann, D., I. Lima Azevedo, P. Ferrão, and J. E. Fernández. 2011. "Residential Electricity Consumption in Portugal: Findings from Top-down and Bottom-up Models." *Energy Policy* 39: 2772–2779.

World Bank. 1999. Impact of Power Sector Reform on the Poor; A Review of Issues and the Literature. https://openknowledge.worldbank.org/bitstream/handle/10986/20317/multi_page.pdf;sequence=1.

A novel approach for the identification of critical nodes and transmission lines for mitigating voltage instability in power networks

Akintunde Samson Alayande ⓘ and Nnamdi Nwulu

Voltage collapse is a major issue combating the effectiveness and optimal operation of modern power systems in recent times. This is a great threat to the security and reliability of modern power systems and, in recent times, it has been a growing concern for power system engineers, researchers and utilities. A prompt identification of transmission lines whose outage could lead to a cascading failure and the sets of nodes where voltage collapse could erupt, during critical outages, is therefore a vital issue for a reliable and secure power system operation. An alternative approach to solving these problems is therefore presented in this paper. The problem is viewed from the graph-theoretical perspective, considering the topological properties of power networks. Application of the fundamental circuit theory is employed and the bus-to-line matrix (BLM) is formulated. This matrix provides clearer insights into the interconnections of the components within the network. This valuable information is captured and used for identifying the critical elements where a suitable location for reactive power support could be placed to avoid voltage collapse of the network. The effectiveness of the proposed approach is tested using a simple 10-bus power network. The results obtained are compared with those obtained from the existing approaches. The results obtained show a strong correlation and agreement between the proposed and the existing approaches.

Introduction

The operation of modern power systems, in which the integrity of the network, in terms of reliability and security will not be compromised, is a complex task. In the quest for higher security and reliability in the operation of modern power systems, modern power systems are now becoming highly interconnected, more complex and heavily loaded. Consequently, modern power systems are now becoming heavily stressed and vulnerable to voltage instability. The aftermath is the frequent occurrence of large-scale blackouts as a result of voltage collapse within the networks. This problem is usually experienced in topologically weak power networks, which are characterized by the shortage of reactive power within the networks (Sikiru et al. 2012). For voltage magnitude at every bus of typical power systems to be maintained and controllable at all times, there is a need to identify suitable locations where reactive power devices could be located within the network (Dhadbanjan and Chintamani 2009; Sikiru et al. 2011). The location of reactive power supports in maintaining acceptable voltage limits within power system networks (Khan and Agnihotri 2013; Sikiru et al. 2011) is therefore an important problem to be solved in modern power systems.

The risk associated with this challenge is inevitable, and could be enormous and disastrous. This is why a power system should be well planned and reliably operated so that customers are not affected by both credible and critical contingencies that could result in a total blackout within the network (Yamashita et al. 2008). Yamashita et al. (2008) discussed extensively the economic impacts of voltage collapse on both developed and developing nations. The risk associated with this impact could be totally avoided or substantially minimized by quick identification of the insecure links and buses where voltage collapse could erupt within the networks (Alayande, Jimoh, and Yusuff 2015; Taylor 2011). In other words, prompt identification of critical elements within a modern power system is highly desirable to control and mitigate the frequent cascading failure and hence voltage collapse whose aftermath could be disastrous. Voltage collapse in power systems has been a serious concern of most power system utilities around the world in recent times. This challenge is usually accompanied by an initial gradual and continuous decrease in the system voltage before it drops rapidly (Adebayo et al. 2018). Other contributing factors include stress power network which results from network congestion or high real power loading of the network, the shortage of reactive power supports in the network, transformer tap changing characteristics, etc. Several large-scale blackouts have been experienced in most countries around the world and this has caused a great concern among power system researchers in recent times (Adebayo et al. 2018; Yamashita et al. 2008).

One of the ways through which the solution can be provided is by reinforcing the network through transmission expansion planning. The main interest of the transmission owners and system operators is to ensure the system is voltage collapse-free but with the minimum investment cost. This approach however requires a huge amount of capital and this may limit its application in future power network. Moreover, transmission expansion planning requires line switching or additions, which greatly adds to the computational complexity of the problem. Consequently, obtaining a solution to the problem, in a large modern power system, is therefore hampered. Hence, the application of this method in solving the problem may be impractical and limited, most especially, in large-sized practical power systems. Hence, there is a need for providing additional or alternative computational tools for providing an efficient solution.

Another solution to the problem is the placement of reactive power supports such as reactors or FACTs devices at various locations within the system where their influence would be most effective. Traditionally, generating stations are usually located far away from load centres due to health and developmental reasons. However, since it is not an easy task to transport reactive power over a long distance, there is a need to find an alternative way of compensating the transmission lines adequately in order to increase the efficiency of the transmission network (Taylor 2011). In order to reduce the overall investment cost of installing VAR sources, it is important to install them in such locations, where the least VAR amount is needed to ensure system voltage security against all severe contingencies. In addition, optimal design and operation of power systems are usually faced with the determination of suitable locations of reactive power devices within the networks. The basic AC power-flow equations on which most of the power system solutions depend are mathematically complex and nonlinear in nature. This makes convergence to the real solution practically impossible in most large-sized and radial power networks. This, however, compounds the challenge of identifying a suitable location for the reactive power resources such as SVC, capacitor banks, etc. Although, this approach is found helpful in resolving the issue, to some extent, its main bottleneck is how to quickly identify suitable locations where the effect of such reactive power supports could be most effective. Conventionally, most authors, in the open literature, have formulated the problem as a nonlinear optimization problem, which is subjected to certain system constraints (Ginarsa, Soeprijanto, and Purnomo 2013). The main challenges in this case are indeed non-convexity of the models and computational difficulties in terms of time and computer memory space. This has posed a lot of challenging tasks and concerns for power system operators and planners. A quick and an efficient methodology for identifying such suitable locations where the reactive power supports could be placed in order to prevent voltage collapse, is therefore, of utmost importance. Moreover, effective identification of locations where reactive power support could be placed requires that all the influential links and nodes that could lead to violations of network integrity in terms of reliability and security be quickly identified and monitored (Ginarsa, Soeprijanto, and Purnomo 2013; Ziari et al. 2010).

This challenge has attracted the attention of several researchers in recent times. For instance, Mustafa, Shareef, and Ahmad (2005) considered an improved methodology, for the allocation of individual generator usage, in a deregulated power system. This approach is, however, time-consuming. More so, it is solely dependent on the solved power-flow, whose solution may not be guaranteed and can only be obtained (if it exists) through iterative means. This problem is more pronounced in a large, ill-conditioned power system. More recently, Mafizul Islam, Alam, and Mohammad Yasin (2013) presented an approach which focused on the influence and management of reactive power on power system stability and security in a restructured or deregulated power system. An approach for solving the reactive power cost allocation problem, based on the power tracing principle, was proposed by De and Goswami (2012). This technique solves the problem of bidirectional reactive power-flow within power networks. The authors also considered reactive power supplied by line charging capacitance as a separate reactive source with their associated reactive load and loss allocated to it. Total system reactive demand and loss were first allocated to all the sources within the system, such as generators, synchronous condensers, capacitors and line charging capacitance. A quadratic reactive cost function was adopted for the generators while all other sources were assumed to have a constant cost per unit MVAR supplied. Khalid et al. (2008) proposed a novel method for identifying the reactive power transfer between generators and load using modified Kirchhoff's laws. The bus admittance matrix is partitioned to decompose the current of the load buses as a function of generator current and voltage based on the solved power-flow solution and the network parameters. The decomposed demand reactive power is obtained from the decomposed currents. This work also creates an appropriate artificial neural network (ANN) for practical 25-bus equivalent power system of south Malaysia for testing the effectiveness of the ANN output compared to that of the modified nodal equation method. The results show that ANN output provides more promising results in terms of accuracy. However, the application of this approach is limited and may not be suitable for large power network due to convergence issue and large computational time associated with it. A graph theory-based approach in conjunction with the power-flow analysis of the network is proposed by Wu, Ni, and Wei (2000) to solve an allocation problem in a deregulated economy. This approach presented promising results when applied to determine the allocation of individual network participants (generators and loads) to the flows on the transmission lines as well as the active power transfer between individual network participants in a deregulated economy. Although the approach has been shown to be effective, its capability in the identification of influential network elements without running the repetitive and time-consuming power-flow analysis, which is iterative-based approach, has not been investigated in a holistic view.

The contributions offered by this approach to the active stream of research are twofold. Firstly, the computational complexity involved in the traditional approaches is avoided in this approach. An attempt is, therefore, made to explore the benefit derived from the inherent structure of power networks and the interconnections that exist between the network components, without the need for solving the time-consuming power-flow equations, in identifying the influential nodes and lines within a power system network. Secondly, this approach is suitable for practical applications due to its simplicity in identifying the influential nodes and lines within the networks. This will help the system operators in making faster decisions for voltage collapse prevention before it actually occurs, thereby protecting the integrity of the networks, most especially, during critical outages.

The remainder of the paper is organized as follows. The next section presents the application of graph theory to a power network and how the advantages inherent in the graph-theoretical approach could be explored in solving power system problems. The description of a case study used to test the effectiveness of the proposed approach is presented in the section thereafter. The penultimate section presents the results, the discussions as well as the comparison of the results obtained with the results obtained using the existing methods while the study is concluded in the final section.

Graph-theoretical approach in power systems

In recent times, graph theory has become an interesting approach to solving most engineering problems. Its applications have been widely deployed and well researched in seeking solutions to problems in most fields of engineering and sciences such as transportation engineering (Alayande 2017). In this paper, the role of graph theory in identifying influential elements within power system networks to prevent voltage collapse within the network, is investigated. In this section, a power network is modelled as a graph in order to actually capture the benefits inherent in the interconnections of the network for identifying critical elements whose outage could lead to voltage collapse within the power system.

Suppose a complex network such as power systems can be represented as a directed graph $G = (V, E)$ with sets of vertices V and a set of edges or loops E. According to the fundamental concepts in graph theory, two vertices i and j are said to be connected, in an undirected graph G, if they are neighbours; a path exists between the two vertices u and v. A connected graph G is one in which all pairs of distinct vertices are connected. A graph G is said to be undirected if none of the edges has orientations. In a graph G, a branch is a link that connects any two vertices u and v and it is referred to as an edge. A set of edges whose removal could lead to separation of a connected graph G into two different components is termed a cutset. Suppose a complex power system is reduced to a simple graph with no multiple loops or edges, then the degree of any node k is the number of neighbours associated with bus k. In other words, the degree of a given node within a network graph indicates the number of nodes which are adjacent to that node.

In a complex network such as power system, which could be described as a graph whose vertices are connected by edges, the following simplifying assumptions hold: the edges are ordered pairs with at most one link joining any two buses and no self-loop exists and all the edges of the graph are not weighted or similar. Considering these assumptions, the elements of the bus-to-line incidence matrix or adjacency matrix $A(G)$ for an undirected graph $G(V, E)$ has its $ij - th$ element equals to 1 if a connection exists between two vertices i and j, and 0 if no connection exists. Mathematically,

$$A(G)_{i,j} \begin{cases} 1, & \text{if } i \text{ is connected to } j \\ 0, & \text{if } i \text{ is not connected to } j \end{cases} \quad (1)$$

For a directed graph,

$$A(G)_{i,j} = \begin{cases} 1, & \text{if line } j \text{ starts at bus } i \\ -1, & \text{if line } j \text{ ends at bus } i \\ 0, & \text{if } j \text{ is not connected to bus } i \end{cases} \quad (2)$$

Hence, the degree of a node can be defined in terms of the adjacency matrix, as the sum of the elements of the corresponding row (or column). The importance of the degree distribution is found in representing the global connectivity of a given network. The node degree can simply be expressed as

$$D(G)_{i,i} = \sum_{j=1}^{n} |a_{ij}| \quad (3)$$

where n is the total number of buses and A is a symmetric matrix.

The Laplacian Matrix, L of any given graph G is defined as

$$L(G)_{i,j} = D(G)_{i,j} - W(G)_{i,j} \quad (4)$$

where W is the weight matrix associated with the vertices of the graph.

Based on the fundamental theorem and the preliminary results obtained by Wu, Ni, and Wei (2000), the following characteristics of BLM are identified:

(1) The summation of all nonzero elements in each column is zero based on Kirchhoff's current law at each node within the network.
(2) The total number of positive 1s in both BOLIM is equal to the number of negative 1s in BILIM.

Algorithm development for the formulation of Bus-to-Line Incidence Matrix

This section presents the three matrices developed for solving the identified problem based on graph theory. These matrices are termed Bus-to-Line Incidence Matrix (BLM), Bus-Inflow Line Incidence Matrix (BILIM) and Bus-Outflow Line Incidence Matrix (BOLIM). The BLM matrix is easily formulated from the graphical representation of the network as shown in Figure 1. The

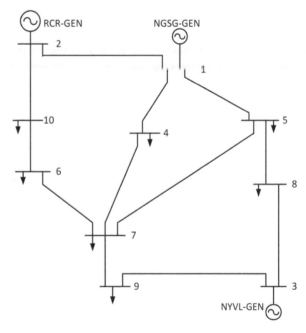

Figure 1: A simple 10-bus network (Alayande, Jimoh, and Yusuff 2015).

BILIM matrix contains the elements of the sinks while the elements of BOLIM correspond to the sources. Based on the upstream and downstream power tracing, pure sources and pure sinks are identified. The rows with zero element values in BOLIM correspond to the pure sink buses while the rows with their elements all zeros in BILIM correspond to the pure source buses. Having identified the pure source buses within the network, corresponding columns (lines) in the BILIM matrix are deleted and the reduced or equivalent network for the network in terms of the two matrices BOLIM and BILIM is easily generated. The process is continued until there is only one pure source or pure sink within the network. During the process, if loop-flow occurs, the bus where it exists supersedes other identified buses where reactive power support needs to be placed.

The procedural steps involved in the algorithm development for the proposed approach are as follows:

(1) Model the network as a weighted directed graph.
(2) Perform arbitrary numbering on the network vertices.
(3) Perform arbitrary numbering of the network edges.
(4) Indicate the directions of power flows on the network edges. For example, power going out of the bus could be assumed to be positive while that coming into the bus would be assumed to be negative.
(5) Compute the elements of BLM and form the BLM matrix.
(6) Formulate the matrices BILIM and BOLIM from the BLM matrix.
(7) Identify the sources and the corresponding sinks within the network.
(8) Perform a matrix reduction on matrix BILIM and its corresponding matrix BOLIM.
(9) Continue the last two steps until the matrices BILIM and BOLIM are no longer reducible.
(10) Identify the critical elements such as weak nodes and weak transmission lines within the network.
(11) End.

Numerical illustration

A numerical example is presented in this section to illustrate the effectiveness of the approach proposed in this paper. For the sake of clear illustration, we consider a simple Southern Indian 10-bus power grid whose single-line diagram is presented in Figure 1. The transmission line data is adapted from Alayande et al. (2015). This system comprises of 12 transmission lines and 10 buses, of which 3 are generators that are geographically located within the network and the remaining 7 are load buses.

Results and discussions

In this section, the benefit derived from the analogy of atomic structures is first explored to determine the structural levels of each node within the network. Table 1 presents the structural levels of each node in a 10-bus network shown in Figure 1 and their corresponding node degrees for the network under consideration. The structural level in the context of this paper refers to the combination of network participants or elements whose structural interconnections form a circular path or shell-like structure. This provides necessary information about the topological distance of each load bus with reference to the generator buses within the network, which could easily be captured without the need to carry out the repetitive and time-consuming power-flow analysis. The generators are assigned zero (0) being the reference nodes and are presented in Table 1. The results obtained for the degree of each network load are also presented in Table 1.

The structural levels of other buses are determined based on their topological interconnections to the nearest generator bus in the network. All the generator buses on structural level 0 are connected on a circular path, which is the innermost circle of Figure 1. The next circular path is the circle that connects all the buses on the structural level 1 while the outermost circle connects all the buses that are on the structural level 2 as presented in Table 1. Also presented in Table 1 are the results obtained for the formulation of BLM matrix for the network from which the BILIM in Table 2 and BOLIM in Table 3 are extracted. The elements of the BLM matrix are obtained by finding the absolute values of all the negative elements contained in BLIM while positive elements of the BLIM matrix are used to form the Bus-Outflow Line Incidence Matrix (BOLIM).

Considering the developed algorithm, pure source buses within the network are identified as buses 1 and 3 while the pure sink buses are identified as buses 8 and 9. The identified pure source buses are therefore deleted in the BILIM, which correspond to the rows 1 and 3. The columns that correspond to pure source buses in BOLIM matrix are those columns whose elements of BOLIM are 1s on rows 1 and 3. Therefore, columns a, b, c, h and i in BOLIM are also deleted.

Table 1: Structural level identification and Bus-to-Line Incidence Matrix (BLM).

Bus no.	Bus type	Structural level	Node degree	Transmission line											
				a	b	c	d	e	f	g	h	i	j	k	m
1	Generator	0	3	1	1	1	0	0	0	0	0	0	0	0	0
2	Generator	0	2	0	0	−1	1	0	0	0	0	0	0	0	0
3	Generator	0	2	0	0	0	0	0	0	0	1	1	0	0	0
4	Load	1	2	−1	0	0	0	0	1	0	0	0	0	0	0
5	Load	1	3	0	−1	0	0	1	0	1	0	0	0	0	0
6	Load	2	2	0	0	0	0	0	0	0	0	0	1	−1	0
7	Load	2	3	0	0	0	0	0	−1	−1	0	0	−1	0	1
8	Load	1	2	0	0	0	0	−1	0	0	−1	0	0	0	0
9	Load	1	2	0	0	0	0	0	0	0	0	−1	0	0	−1
10	Load	1	2	0	0	0	−1	0	0	0	0	0	0	1	0

Table 2: Bus-Inflow Line Incidence Matrix (BILIM).

Bus no.	Transmission line											
	a	b	c	d	e	f	g	h	i	j	k	m
1	0	0	0	0	0	0	0	0	0	0	0	0
2	0	0	1	0	0	0	0	0	0	0	0	0
3	0	0	0	0	0	0	0	0	0	0	0	0
4	1	0	0	0	0	0	0	0	0	0	0	0
5	0	1	0	0	0	0	0	0	0	0	0	0
6	0	0	0	0	0	0	0	0	0	0	1	0
7	0	0	0	0	0	1	1	0	0	1	0	0
8	0	0	0	0	1	0	0	1	0	0	0	0
9	0	0	0	0	0	0	0	0	1	0	0	1
10	0	0	0	1	0	0	0	0	0	0	0	0

Table 3: Bus-Outflow Line Incidence Matrix (BOLIM).

Bus no	Transmission line											
	a	b	c	d	e	f	g	h	i	j	k	m
1	1	1	1	0	0	0	0	0	0	0	0	0
2	0	0	0	1	0	0	0	0	0	0	0	0
3	0	0	0	0	0	0	0	1	1	0	0	0
4	0	0	0	0	0	1	0	0	0	0	0	0
5	0	0	0	0	1	0	1	0	0	0	0	0
6	0	0	0	0	0	0	0	0	0	1	0	0
7	0	0	0	0	0	0	0	0	0	0	0	1
8	0	0	0	0	0	0	0	0	0	0	0	0
9	0	0	0	0	0	0	0	0	0	0	0	0
10	0	0	0	0	0	0	0	0	0	0	1	0

Table 4: Bus-Inflow Line Incidence Matrix (BILIM).

Bus no	Transmission line						
	d	e	f	g	j	k	m
2	0	0	0	0	0	0	0
4	0	0	0	0	0	0	0
5	0	0	0	0	0	0	0
6	0	0	0	0	0	1	0
7	0	0	1	1	1	0	0
8	0	1	0	0	0	0	0
9	0	0	0	0	0	0	1
10	1	0	0	0	0	0	0

Table 6: Bus-Inflow Line Incidence Matrix (BILIM).

Bus no	Transmission line		
	j	k	m
6	0	1	0
7	0	0	0
8	0	0	0
9	0	0	1
10	1	0	0

The results

Based on the results presented in Table 5, pure source buses within the network are identified as 2, 4 and 5 while the pure sink buses are buses 8 and 9. Therefore, the identified pure source buses and their correspondence columns are deleted. These correspond to the rows 2, 4 and 5 in Table 4 and columns d, e and f in Table 5. The

results obtained for the reduced BILIM and BOLIM matrices are presented in Tables 6 and 7 respectively.

In a similar manner, pure source buses within the reduced network are buses 7 and 8 while pure sink bus is bus 8. Therefore, from Table 6, the pure source buses are deleted which correspond to rows 7 and 8 and their corresponding columns in Table 7 are also deleted which correspond to line m. The result obtained for the reduced BILIM and BOLIM matrices, after the deletion, are presented in Tables 8 and 9 respectively. It can be seen from Tables 8 and 9 that bus 9 corresponds to a pure source bus and pure sink. Therefore, row 9 in Tables 9

Table 5: Bus-Outflow Line Incidence Matrix (BOLIM).

Bus no	Transmission line						
	d	e	f	g	j	k	m
2	1	0	0	0	0	0	0
4	0	0	1	0	0	0	0
5	0	1	0	1	0	0	0
6	0	0	0	0	1	0	0
7	0	0	0	0	0	0	1
8	0	0	0	0	0	0	0
9	0	0	0	0	0	0	0
10	0	0	0	0	0	1	0

Table 7: Bus-Outflow Line Incidence Matrix (BOLIM).

Bus no	Transmission line		
	j	k	m
6	1	0	0
7	0	0	1
8	0	0	0
9	0	0	0
10	0	1	0

Table 8: Bus-Inflow Line Incidence Matrix (BILIM).

Bus no	Transmission line	
	j	k
6	0	1
9	0	0
10	1	0

Table 9: Bus-Outflow Line Incidence Matrix (BOLIM).

Bus no	Transmission line	
	j	k
6	1	0
9	0	0
10	0	1

Table 10: Bus-Inflow Line Incidence Matrix (BILIM).

Bus	Transmission line	
	j	k
6	0	1
10	1	0

Table 11: Bus-Outflow Line Incidence Matrix (BOLIM).

Bus	Transmission line	
	j	k
6	1	0
10	0	1

and 10 is deleted. The results obtained for the reduced equivalent BILIM and BOLIM matrices are presented in Tables 10 and 11.

It can be seen that no pure source or pure sink bus exists in Tables 10 and 11. This shows the end of line and bus deletion within the network. Furthermore, the transmission lines not deleted, as can be seen in Tables 10 and 11, are lines j and k while the buses that are not deleted until the end of the process are buses 6 and 10. By employing the electrical characteristics of the network, the most influential node and the weakest transmission lines can be easily identified. Hence, the most suitable and appropriate location where the reactive support could be placed can easily be identified. Line j corresponds to the transmission line joining buses 6 and 7 with an electrical distance of 248.7593 while line k

represents the transmission line joining buses 6 and 10 with an electrical distance of 14.6716.

Therefore, bus 6 is identified as the most influential bus within the network, where the location of a reactor will have an influence on the operation of the system. Also, line j, which connects buses 6 and 7, is the most critical transmission line within the network under study. This information about transmission lines within the network could be of utmost importance during the outage of a critical transmission line.

Contingency analysis

In this section, we carry out a contingency analysis based on N-1 criterion. This is necessary in order to further confirm and affirm the results obtained from the proposed approach. The influence of a critical outage of the identified transmission line, which connects buses 6 and 7, based on the N-1 criterion, on the integrity of the network is investigated. Power-flow analysis is carried out to further test the efficiency of the method proposed in this paper. Table 12 presents the results obtained from the power-flow solution as well as the eigenvalue and eigenvector analyses. From Table 12, for the base case, it is seen that the voltage magnitudes at buses 6 and 7 are out of the prescribed limits of 0.95–1.05 pu and the active power loss of 18.573 MW is obtained. When the most influential transmission line (the line connecting buses 6 and 7) is outaged, the results show that buses 4, 7 and 9 are outside the prescribed limits and the active power loss within the network increases to 19.960 MW. However, by injecting reactive power at the identified weak bus 6, the results show a considerable improvement in the voltage profile and the active power loss is significantly reduced to 17.35 MW.

Based on the eigenvalue and eigenvector analyses (Sikiru et al. 2012), the smallest eigenvalues for both normal and outage operations are found to be associated with bus 6 with a value of 0.0000 as presented in the last two columns of Table 13. This further confirms that bus 6 is the weakest bus in the network. Also, elements of the eigenvector associated with this smallest eigenvalue, for all the buses, are found to be equal to the value of 0.380 as depicted in Figure 2. This indicates that the network is topologically weak and needs to be reinforced by injection of reactive power at the weakest bus, namely bus 6.

Table 12: Results obtained for a critical outage on line 6–7 in a simple 10-bus network.

Bus No	Bus type	Voltage magnitude (pu)			Eigenvalue	
		Base case	During outage	With VAR injection at bus 6	After outage	Before outage
1	Generator	1.0000	1.0000	1.0000	-	-
2	Generator	1.0500	1.0500	1.0500	-	-
3	Generator	1.0300	1.0300	1.0300	-	-
4	Load	1.0481	1.0672	1.0443	100.1787	540.5414
5	Load	1.0367	1.0466	1.0281	63.1080	81.6696
6	Load	1.0544	0.0008	1.0248	0.0000	0.00000
7	Load	1.0547	1.0705	1.0289	6.2914	58.2337
8	Load	1.0171	1.0337	1.0130	6.2914	41.0213
9	Load	1.0353	1.0548	1.0264	6.2914	31.6160
10	Load	1.031	1.0478	1.0494	36.1679	23.1600
Active Power Loss		18.573 MW	19.960 MW	17.35 MW		

Table 13: Comparison of approaches for weakest load bus identification.

		Ranking		
Load bus no	Proposed approach	Coupling strength index-based approach (Alayande, Jimoh, and Yusuff 2015)	Eigenvalue-based approach (Sikiru et al. 2011)	T-index-based approach (Dhadbanjan and Chintamani 2009)
4	7th	7th	7th	5th
5	6th	5th	6th	3rd
6	**1st**	**1st**	**1st**	**1st**
7	4th	2nd	5th	2nd
8	5th	6th	4th	7th
9	3rd	4th	3rd	4th
10	2nd	3rd	2nd	6th

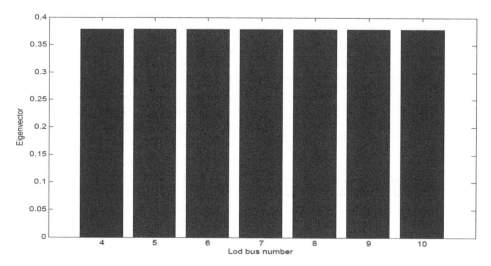

Figure 2. Elements of the eigenvector associated with the weakest bus in a 10-bus network.

Comparison of approaches

Table 13 presents a comparison of the results obtained from the existing approaches with those obtained using the proposed approach.

The results obtained in this paper are in strong agreement with the results obtained using inherent structural characteristics as presented by Alayande, Jimoh, and Yusuff (2015). The results obtained using the proposed approach also compare well with the existing approach of eigenvalue decomposition presented by Sikiru et al. (2011) and the T-index approach proposed by Dhadbanjan and Chintamani (2009). All three existing methods are in agreement with the most suitable location identified by the proposed method where the influence of reactive power support within the network will be greatly appreciated.

The main benefit of the proposed approach over the existing approaches is that this approach can easily identify both influential nodes and transmission lines without the need to perform power-flow analysis. The issues associated with the slack bus identification are therefore avoided. Moreover, a considerable amount of time is therefore saved as there is no need for performing the time-consuming power-flow analysis before the solution can be arrived at, contrary to the solution approach presented by Wu, Ni, and Wei (2000). Furthermore, it does not require partitioning analysis, which is highly required for the solution to be obtained in the approach presented by Alayande, Jimoh, and Yusuff (2015).

We also compare the results obtained based on our approach with those obtained through the use of the coupling strength index-based approach (Alayande, Jimoh, and Yusuff 2015) for the identification of the weakest transmission line. The results of the comparison are presented in Table 14. From the results presented in Table 14, it can be seen that both methods identified the line connecting buses 6 and 7 as the weakest transmission line. The main benefit of the proposed approach over the existing approaches is that this approach easily identifies both influential nodes and transmission lines without the need to perform power-flow analysis. A considerable amount of time is therefore saved as there is no need for performing the time-consuming power-flow analysis before the solution can be arrived at, contrary to the solution approach presented by Wu, Ni, and Wei (2000).

Table 14: Comparison of approaches for critical transmission line identification within Load-Load attraction region.

	Ranking	
Bus No – Bus No	Proposed approach	Coupling strength index-based approach
4–7	4th	5th
5–8	3rd	4th
6–7	**1st**	**1st**
6–10	2nd	6th
7–9	6th	2nd
5–7	5th	3rd

Furthermore, it does not require partitioning analysis, which is highly required for the solution to be obtained in the approach presented by Alayande, Jimoh, and Yusuff (2015).

Conclusion

In this paper, identification of a suitable location of a reactor where the operation of power systems could be improved based on graph theory has been considered. The network matrices termed BLIM, BILIM and BOLIM are developed, from graph theory, for identifying the influential nodes and links within power grids. A Southern Indian 10-bus power grid is used as the case study. The results obtained compared well with that obtained by the existing methods as documented in the literature. The comparison shows a strong agreement with the existing results obtained by the past researchers using different methods. This confirms the accuracy of the proposed approach. Based on the results obtained, the proposed approach could be helpful as an alternative algorithm for locating a suitable node, where a reactive power support could be placed for efficient operation of power systems. Moreover, by this method, prompt identification of weak transmission lines and critical buses whose outage could have adverse effect on the stability of the network is made easy. It could also be useful to the system operators in identifying an influential transmission line for power rerouting during an outage of a critical transmission line within the network. The information, if provided, could safeguard the network from collapsing.

Disclosure statement

No potential conflict of interest was reported by the authors.

ORCID

Akintunde Samson Alayande http://orcid.org/0000-0003-1503-2171

References

Adebayo, I. G., A. A. Jimoh, A. A. Yusuff, and Y. Sun. 2018. "Alternative Method for the Identification of Critical Nodes Leading to Voltage Instability in a Power System." *African Journal of Science, Technology, Innovation and Development* 10: 323–333. doi:10.1080/20421338.2018.1461967.

Alayande, A. S. 2017. "Solving Power System Problems Based on Network Structural Characteristics." D.Tech. Thesis, Department of Electrical Engineerig, Tshwane University of Technology, Pretoria, South Africa.

Alayande, A. S., A. A. Jimoh, and A. A. Yusuff. 2015. "Identification of Critical Buses and Weak Transmission Lines Using Inherent Structural Characteristics Theory." IEEE power and energy engineering conference (APPEEC), 2015 IEEE PES Asia-Pacific, 1–5, Brisbane,

Australia, IEEExplore, November 15–18. doi:10.1109/APPEEC.2015.7380974.

De, M., and S. K. Goswami. 2012. "Reactive Power Cost Allocation by Power Tracing Based Method." *Energy Conversion and Management* 64: 43–51.

Dhadbanjan, T., and V. Chintamani. 2009. "Evaluation of Suitable Locations for Generation Expansion in Restructured Power Systems: A Novel Concept of T-Index." *International Journal of Emerging Electric Power Systems* 10 (1): 1–25. doi:10.2202/1553-779X.2023.

Ginarsa, I. M., A. Soeprijanto, and M. H. Purnomo. 2013. "Controlling Chaos and Voltage Collapse Using an ANFIS-Based Composite Controller-Static Var Compensator in Power Systems." *International Journal of Electrical Power & Energy Systems* 46: 79–88.

Khalid, S. N., M. W. Mustafa, H. Shareef, A. Khairuddin, A. Kalam, and O. A. Maungthan. 2008. "A Novel Reactive Power Transfer Allocation Method with the Application of Artificial Neural Network." 2008 IEEE Australasian university power engineering conference (AUPEC), Sydney, NSW, Australia.

Khan, B., and G. Agnihotri. 2013. "Optimal Transmission Network Usage and Loss Allocation Using Matrices Methodology and Cooperative Game Theory." *International Journal of Electrical, Robotics, Electronics and Communications Engineering, World Academy of Science, Engineering and Technology* 7 (2): 142–149.

Mafizul Islam, Sk, M. Alam, and Sk Mohammad Yasin. 2013, July. "A New Approach of Reactive Loss Allocation in Deregulated Power System." *International Journal of Advanced Research in Electrical, Electronics and Instrumentation Engineering* 2 (7): 2948–2954.

Mustafa, M. W., H. Shareef, and M. R. Ahmad. 2005. "An Improved Usage Allocation Method for Deregulated Transmission Systems." IEEE 2005 international power engineering conference, Singapore.

Sikiru, T. H., A. A. Jimoh, and J. T. Agee. 2011. "Optimal Location of Network Devices Using a Novel Inherent Network Topology Based Technique." IEEE Africon 2011 – the falls resort and conference centre, Livingstone, Zambia, 1–4, September 13–15.

Sikiru, T. H., A. A. Jimoh, Y. Hamam, J. T. Agee, and R. Ceschi. 2012. "Classification of Networks Based on Inherent Structural Characteristics." Sixth IEEE/PES Transmission and Distribution: Latin America Conference and Exposition (T&D-LA), Montevideo, 1–6. doi:10.1109/TDC-LA.2012.6319055.

Taylor, J. A. 2011, June. "Conic Optimization of Electric Power Systems, in Partial Fulfilment of the Requirements for the Degree of Doctor of Philosophy." Massachusetts Institute of Technology.

Wu, F. F., Y. Ni, and P. Wei. 2000. "Power Transfer Allocation for Open Access Using Graph Theory—Fundamentals and Applications in Systems Without Loopflow." *IEEE Transactions On Power Systems* 15 (3): 923–929.

Yamashita, K., S. Joo, J. Li, P. Zhang, and C. Liu. 2008. "Analysis, Control, and Economic Impact Assessment of Major Blackout Events." *European Transactions on Electrical Power* 18: 854–871.

Ziari, I., G. Ledwich, A. Ghosh, D. Cornforth, and M. Wishart. 2010. "Optimal Allocation and Sizing of Capacitors to Minimize the Transmission Line Loss and to Improve the Voltage Profile." *Journal of Computers and Mathematics with Applications* 60: 1003–1013.

Design, construction and mathematical modelling of the performance of a biogas digester for a family in the Eastern Cape province, South Africa

Patrick Mukumba, Golden Makaka, Sampson Mamphweli and Peace-maker Masukume

Currently, South Africa is experiencing electricity blackouts as result of an energy shortage. Biogas can be a solution to South Africa's energy needs, especially in the rural areas of the Eastern Cape province that have plenty of biogas substrates from cattle, donkeys, goats, sheep and chicken. The purpose of this paper is to design and construct a 1 m³ family biogas digester and model its performance using donkey dung as a biogas substrate. The installation of this type digester in rural communities of South Africa has economic, social and environmental advantages. The digestate from the digester is a valuable soil fertilizer, rich in nitrogen, phosphorus, potassium and micronutrients, which can be applied on soils. The mathematical model equation developed is highly beneficial in the rural areas of the Eastern Cape province of South Africa where many donkeys are kept. Currently, no mathematical models have been developed for optimum methane yield from the named substrate.

Introduction

Anaerobic digestion is processes whereby microorganisms break down biodegradable material such as donkey dung in the complete absence of oxygen to form biogas. The biogas produced can be used for electricity generation and as a fuel for vehicles. South Africa is the most industrialized country in Africa and is highly dependent on conventional fuels such as coal and oil. This makes the country lead in greenhouse gas emissions (Naidoo 2011). A number of steps have been taken to reduce these greenhouse gas emissions. One step taken was the introduction of anaerobic digesters in rural areas to produce biogas. Biogas production is a suitable technology used for treatment of organic wastes such as municipal wastes and the production of energy from the combustion of biogas (Lema and Omil 2001; Lettinga 2001; McCarty 2001). Anaerobic digestion is the production of biogas, mainly methane, from organic wastes in the complete absence of oxygen by anaerobic microbes such as acidogens, acetogens and methanogens.

The main factors in the production of biogas involve temperature inside the digester, retention time (RT), agitation, working pressure of the digester, fermentation medium pH, volatile fatty acids (VFA) and sublayer composition (Dobre, Nicolae, and Matei 2014).

The temperature for the anaerobic process can be classified into three conditions which include psychrophilic (between 10°C and 20°C), mesophilic (between 22°C and 40°C) and thermophilic (between 50°C and 60°C) (Cheng 2009; Vintilă et al. 2010).

Biogas technology is an appropriate technology for the recovery of biogas (Etuwe, Momoh, and Iyagba 2016). The biogas production process produces less greenhouse gas than waste treatment processes (Walker, Charles, and Cord-Ruwisch 2009) and landfilling (Lou and Nair 2009). Currently, there are about 200 biogas digesters in operation in South Africa, of which 90% are of the small-scale domestic variety (Tiepelt 2013). The 200 biogas digesters, installed mainly by non-governmental organizations, cannot compare numerically with the vast numbers in both India with 12 million and China with 17 million biogas digesters (DOE 2015). The most common biogas digester types in South Africa are PVC digesters, fixed dome digesters and plastic bag (balloon) digesters. PVC biogas digesters have advantages that include no mechanical wear, can be welded and joined easily, are light in weight, have both high mechanical strength and toughness, and are durable, non-toxic and cost effective.

Fixed dome digesters are usually built underground (Santerre and Smith 1982). The size of the biogas digester depends on the location, number of households, and the amount of feedstock available daily (Rajendran, Aslanzadeh, and Taherzadeh 2012). The advantages of fixed dome biogas digesters include low construction costs, being corrosion free because they are not made from steel, a longer lifespan than other digester types, minimized temperature fluctuations due to underground construction and creation of local employment for construction, feeding and maintenance of the digesters. However, fixed dome digesters have a number of disadvantages such as fluctuation of biogas pressure, low biogas digester temperatures and poor agitation leading to low biogas yield; also, they are not easy to the clean, and require supervision during construction and skilled builders for the construction (Sharma and Pellizzi 1991; Nijaguna 2002).

Plastic bag biogas digesters have advantages that include the following: they are cheap to buy, ease of transportation to installation sites, easy to clean, high biogas temperatures during summer leading to high biogas yield and easy to empty. However, their disadvantages include low gas pressure (as a result, biogas pumps are used), poor agitation, short lifespan, great susceptibility to mechanical damage. Further, the digesters are not locally made, it is not easy to repair the plastic and it is difficult to insulate to prevent temperature fluctuations.

The following problems and challenges of biogas technology in South Africa were noted (Mukumba et al. 2016b):

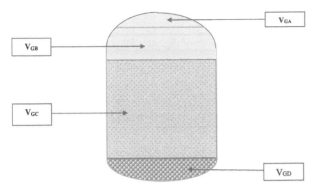

Figure 1: Cross-section of the designed cylindrical biogas digester. *KEY* Volume of gas collecting chamber = V_{GA}; Volume of gas storage chamber = V_{GB}; Volume of fermentation chamber = V_{GC}; The volume of the sludge layer = V_{GD}; Total volume of the digester, $V = V_{GA} + V_{GB} + V_{GC} + V_{GD}$.

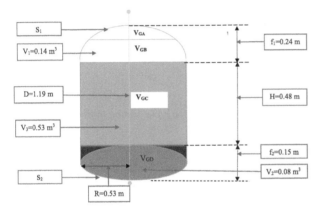

Figure 2: Geometrical dimensions of the cylindrical shaped biogas digester body (Mukumba, Makaka, and Mamphweli 2017).

- High initial investment cost for constructing biogas digesters
- Unavailability of biogas substrates
- Limited research on biogas technology
- Biogas pilot phase failure
- No awareness programme
- Inadequate expertise for construction and maintenance
- Hydrogen sulphide and carbon dioxide in biogas
- Low efficiency of biogas
- Low ESKOM electricity costs
- Low application of biogas

Some of the solutions to the biogas challenges and problems are co-digestion of substrates, improving the calorific value of biogas, extensive research work in biogas technology and adequate education of biogas users (Mukumba, Makaka, and Mamphweli 2018).

There is a need to come up with new design models of biogas digesters which are suited to their environmental conditions, and are easy to use, agitate and clean. They should be easy to construct and feed. Hence, the aim of this paper was to design, install and model the performance a donkey dung fed digester. Currently, there are no model equations developed for mono-digestion of donkey dung. The model equation is important for the Eastern Cape province of South Africa since it has plenty of donkey dung.

Methodology
Designing a batch biogas digester
Design parameter
The following aspects were considered during the design process of the batch biogas digester: durability, air tightness, availability of local materials for the construction process and easy operation (Mukumba, Makaka, and Mamphweli 2017). The design parameters included total solids, suitable temperature, pH and carbon/nitrogen ratio and retention time.

Cross-section of a cylindrical batch biogas digester
A cylindrical design for the batch biogas digester was chosen because it is easy to design and construct, agitate, feed and insulate. It is also easy to remove slurry after every hydraulic retention period. Figure 1 shows the cross-section of a batch biogas digester.

KEY
Volume of gas collecting chamber = V_{GA}
Volume of gas storage chamber = V_{GB}
Volume of fermentation chamber = V_{GC}
The volume of the sludge layer = V_{GD}
Total volume of the digester, $V = V_{GA} + V_{GB} + V_{GC} + V_{GD}$

Figure 2 shows the cylindrical batch digester body with all calculated values indicated. Table 1 shows calculated volume and geometrical dimensions of the batch biogas digester.

Volume calculation of digester and hydraulic chamber
The volume of the digester chamber was calculated using assumptions for volume and geometrical dimensions. In the calculations, a retention period of 30 days and a mesophilic temperature of 30°C were used, as mentioned earlier. Table 2 shows the calculated values of the digester chamber from the geometrical and volume assumptions.

Table 1: Assumptions for volume and geometrical dimensions (REEIN, 2012).

Assumptions: For volume	For geometrical dimensions
$V_{GD} = 5\%V$	$D = 1.3078 \times V^{1/3}$
$V_{GB} + V_{GC} = 80\%V$	$V_1 = 0.0827\ D^3$
$V_{GD} = 15\%V$	$V_2 = 0.05011\ D^3$
$V_{GB} = 0.5\ [V_{GB} + V_{GC} + V_{GD}]\ k$	$V_3 = 0.3142\ D^3$
$k = 0.4 m^3/day$	$R_1 = 0.725\ D;\ R_2 = 1.0625\ D$
$V = 0.8$	$f_1 = D/5;\ f_2 = D/8$
	$S_1 = 0.911\ D_2;\ S_2 = 0.8345\ D^2$

Table 2: Volume calculation of digester and hydraulic chamber.

Calculated values of digestive chamber	Calculated values of digester chamber
$f_1 = D/5 = 0.24$ m	$D = 1.3078V^{1/3} = 1.19$ m
$R_1 = 0.725D = 0.86$ m	$f_2 = D/8 = 0.15$ m
$V_1 = 0.0827D3 = 0.14$ m^3	$R_2 = 1.0625D = 1.26$ m
$V_3 = 0.3142D^3 = 0.53$ m^3	$V_{GA} = 0.05$ V $= 0.04$ m^3
$V_{GD} + V_{GC} = 0.6$ m^3	$H = (4 \times 0.3142D^3)/3.14D^2 = 0.48$ m

Table 3: Parameters and equation for calculating the working volume of the batch biogas digester.

Parameters	Magnitude
Temperature	30°C
Retention time (HRT)	30 days
Concentration of TS	8%
Working volume $(V_{GD} + V_{GC}) - Q.HRT$	0.6 m^3

Table 3 shows a summary of the parameters and calculation of the working volume of the batch biogas digester.

The cylindrical batch biogas digester construction processes

The success or failure of any biogas digester mainly depends upon the quality of the construction work. The construction of the batch biogas digester involved several steps.

Selection of construction material for the biogas digester

The materials used in the digester construction were cement, sand, aggregate, water and bricks. High quality Portland cement was used in the digester construction. The sand used for construction purposes was cleaned in order to increase the digester strength. Water was mainly used for preparing the mortar for masonry work, concreting work and plastering. It was also used to soak the clinker bricks before using them. The clinker bricks used were of the best quality, with high compressive strength and locally available. They are water-resistant, durable and have high thermal conductivity. The aggregates used for masonry work were clean, strong and of good quality. In addition, high quality epoxy paint was used for painting the inside of the biogas digester because it has a high waterproofing features and low thermal conductivity of 0.30 W/(m.K). Locally available sawdust was selected for insulation because of its low thermal conductivity of 0.08 W/(m.K) (Mukumba, Makaka, and Mamphweli 2017).

Site selection and layout

For optimum operation of the biogas digester, a site close to a water supply was chosen. This site was also suitable because the land slopes gently (to the north), thus allowing easy water drainage. After selection of the site, the centre of the digester to be constructed was marked on the ground surface with a steel rod. A string cord of length 530 mm was attached to the fixed steel rod stuck underground. The other end of the cord was attached to a marking device (peg). The circumference of the batch biogas digester was marked by rotating the end of the cord in a circular fashion. The cord extended to a length of 530 mm to the

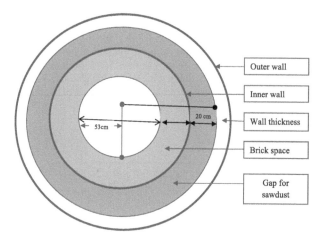

Figure 3: Sketch showing the cross section of the layout of the digester.

second circumference. Figure 3 shows the sketch layout of the biogas digester. Once site layout was complete it was reviewed to ensure best site selection.

Excavation

A cylindrical pit 500 mm deep was dug. The excavated soil was placed two metres away from the edge of the dug pit to ensure that there no soil could fall inside the pit during the construction process. Furthermore, the pit bottom was levelled.

Construction of the foundation

The cylindrical digester pit, 500 mm deep and with a radius of 540 mm, was filled with concrete. Figure 4 shows the sketch (depth and width) of the digester pit. The ratio of cement, sand and stone aggregates used for the mixture were 1: 2: 3. The concrete was rammed to increase strength and was left for seven days to allow settling. Water was poured on the concrete slab twice daily during the seven days, to avoid cracking. The concrete slab was first covered with a black plastic sheet

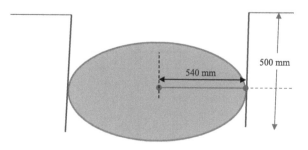

Figure 4: Sketch of the depth and width of the digester foundation/base.

Figure 5: First and second brickwork after the foundation.

(damp course) in order to prevent the transfer of moisture from the ground to the slurry to keep the water/solids ratio to the measured one. The first course of bricks with mortar was aligned on top of the plastic. The cement to sand ratio for the mortar was 1: 3.

A 700 mm high double wall was constructed. The structure was reinforced with a 230 mm brick force after every two courses. The openings for the mechanical stirrer, thermocouples and pressure gauge (for gas pressure) were left in the wall of the digester. Figure 5 shows a construction step of the batch biogas digester.

Construction of the dome

The dome of the digester was constructed separately from the main body of the digester. The construction of the dome involved several steps.

Step 1: Two sheets of chip wood were glued together to form a single wooden sheet of 3000 × 3000 mm. The wooden sheet was placed on a piece levelled ground.

Step 2: The wooden sheet was covered with a sheet of black plastic. A circle of radius 530 mm was marked on the plastic sheet with a pencil and a second circle was also marked. Figure 6 shows a clear sketch diagram of wooden sheet, plastic sheet, and the first and second circumferences.

Step 3: Stone aggregate was placed in the inter circle on the plastic sheet to form a cone shaped structure with a circular base (diameter) of 1060 mm as shown in Figure 6. The stone aggregate had a height of 240 mm from the surface of the wooden sheet. The stone aggregate dome was covered with

a black plastic sheet to the separate stone aggregate from the reinforced concrete.

Step 4: The curved structure of the biogas digester was designed using iron bars. The bottom part of the structure was circular with a diameter of 1100 mm. It was made up of 12 mm rounded iron bars. The iron bars were aligned on the structure to ensure that there were small spaces between the wires. The 230 mm brick force was also part of the wire mesh forming the curved dome structure. The dome wire structure had two layers: the inner (lower) and outer (upper) layers. The inner layer was 240 mm from the surface, while the outer layer was 260 mm from the surface. The cone wire mesh was fitted snugly onto the dome aggregate. A 400 mm slurry inlet pipe with a diameter of 110 mm was inserted into the wire mesh. In addition, another 150 mm slurry inlet pipe was inserted inside the wire mesh. The rest of the pipe was outside the dome. Similarly, a galvanized gas pipe with a thickness of 15 mm and with a length of 300 mm was also inserted. However, the 150 mm of the pipe was within the wire mesh. Furthermore, a black sheet on the surface of dome aggregate separated the aggregate underneath from the concrete dome wire mesh. Figure 7 shows a detailed sketch of the digester dome with gas and slurry pipes.

Step 5: The pouring of the concrete onto the wire mesh dome was the next step. The concrete mixture was prepared with cement, sand and aggregate in the ratio of 1: 2: 3. The concrete was also rammed. It was ensured that all the dome wires were embedded into the concrete by adding more concrete. Water was later poured onto the concrete dome. This was done on successive 10 days so as to strengthen the dome by preventing the formation of cracks.

Inside plastering and insertion of a mechanical stirrer

The first layer of plaster was done before flooring the biogas digester. The designed mechanical stirrer was fitted before the outside plastering of the digester. The mechanical stirrer was designed in accordance with the diameter and height of the batch digester. The blades of the mechanical stirrer were 300 mm in length to ensure that a homogenous mixture is made during stirring. The device was painted with red oxide to prevent rust. The handle was 300 mm long to improve the mechanical

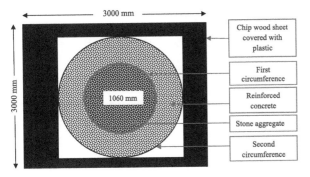

Figure 6: Sketch showing wooden sheet, positions of stone aggregate and reinforced concrete.

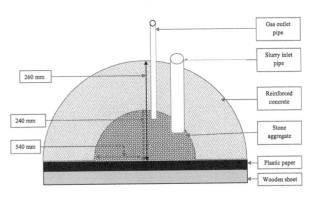

Figure 7: Sketch diagram of the digester dome.

Figure 8: Shows inside epoxy painted floor and walls.

advantage of the system. The mechanical stirrer was hollowed to increase strength mass ratio and for easy stirring of the slurry.

Plastering the outside of the digester and painting

A 110 mm slurry outlet pipe was fitted before outside plastering and painting of the batch digester. Three plastering layers were applied to avoid gas leakages. The thickness of the plaster was 100 mm. The cement to sand ratio for the mortar was 1:3. Epoxy paint was applied on the floor and on the inside walls of the biogas digester to minimize water absorption. Figure 8 shows the epoxy painted digester.

Second wall construction and insulation of the biogas digester

A single outer brick wall was constructed to increase thermal insulation. The separation gap for the insulation for the two separate walls is shown in Figure 9.

The sawdust insulated digester is shown in Figure 10. After plastering of the second wall, dry sawdust was placed between the two walls as shown in the same figure.

Figure 9: A two wall batch digester.

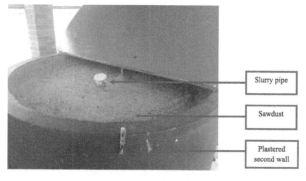

Figure 10: Sawdust insulated biogas digester.

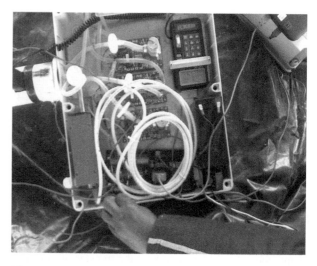

Figure 11: The data aquisition system.

Feeeding the biogas digester

The batch biogas digester was fed with fresh donkey dung. The following parameters were measured in the donkey dung: total alkalinity (T_A), volatile solids (VS), total solids (TS), pH, calorific value (CV), chemical oxygen demand (COD) and ammonium-nitrogen (NH_4-N). All the analytical determinations were performed according to the standard methods for examination of water and wastewater (APHA 2005).

Biogas analysis

A biogas analyzer measured the composition of biogas from the batch biogas digester. The analyzer consisted of a non-dispersive infrared sensor for sensing methane and a carbon dioxide and palladium/nickel (Pd/Ni) sensor for sensing hydrogen and hydrogen sulphide. A CR 1000 data logger, powdered by a 12 V DC battery, stored the data for the composition of the biogas. In addition, a Type K thermocouple thermometer measured slurry and ambient temperatures. The thermocouple was also connected to the data logger. Furthermore, the biogas production rate was measured by a flow metre. Figure 11 shows the data acquisition system that was developed for measurement of biogas composition and temperature.

Results and discussions

Table 4 shows characteristics of donkey dung. The biogas composition had an average methane yield of 55% (Mukumba et al. 2016a).

A mathematical model was developed for methane yield of donkey dung as a function of the following inputs: pH, COD, NH_4-N, digester temperature (T_D) and total alkalinity (T_A).

Table 4: Substrate characteristics for donkey dung (Mukumba et al. 2016).

Parameter	Unit	Donkey dung
Total solids	mg/L	198778.83
Volatile solids	mg/L	144189.99
Total alkalinity	mg/L	6276 - 6343
Ammonium-nitrogen	mg/L	940 - 1223
Calorific value	MJ/g	29.83
Volatile solid /total solid %	%	72.54

Mathematical modelling of the performnce of the batch biogas digester fed with donkey dung

The mathematical model for the experimental results of donkey dung was developed using the MATLAB software package as opposed to other software packages such as Microsoft Excel or SPSS. MATLAB software was used in this research because of its excellent capabilities in mathematical modelling and optimization. The rich statistical toolboxes of MATLAB allow it to perform almost every aspect of mathematical and statistical engineering applications. In addition, the MATLAB can perform:

- ANOVA one test.
- Relief F test-used to rank input variables with respect to the importance or weight(s) of the output.
- A simulation algorithm can be built with the Simulink environment to visualize real-time performance of the system. The MATLAB Simulink is not a customized package but a generalized package. This means simulations from every field of research can be performed.
- Optimization of system performance can be easily performed with the MATLAB optimization tools.

Since there were several input variables for the biogas yield, a multiple regression was used. The model for the multiple linear regression (Mukumba, Makaka, and Mamphweli 2018) is given by:

$$Y_m = \alpha_0 + \alpha_1 x_1 + \alpha_2 x_2 + \alpha_3 x_3 + \alpha_4 x_4 + \ldots \ldots$$
$$+ \ldots \ldots \alpha_n x_n \tag{1}$$

where α_1 to α_n are termed the regression coefficients and x_1 to x_n are called predictor variables of Y_m. The constant, α_0, improves accuracy of results or helps to achieve best fit. In addition, α_0 caters for the factors affecting methane yield other than the ones used in the development of the model. A multiple regression model was developed for methane yield as a function of the following inputs: pH, COD, NH_4-N, digester temperature (T_D) and total alkalinity (T_A). The general theoretical equation for the optimized model (Mukumba, Makaka, and Mamphweli 2018) for the methane yield is expressed as:

$$Y_m = \alpha_0 + \alpha_1(pH) + \alpha_2(COD) + \alpha_3(NH_4 - N)$$
$$+ \alpha_4(T_D) + \alpha_5(T_A) \tag{2}$$

where Y_m is the output (methane yield) and α is a scaling coefficient.

Optimization was done to minimize errors embedded in the experiments. For optimization the MATLAB tool was used. The tool is used for optimizing the methane yield using the solver constraint linear least square multiple regression model to predict the output from the various input variables (Mukumba, Makaka, and Mamphweli 2018). All the α coefficients in the model equation were used for the constraints. From the final values obtained for each set of input variables corresponding to the measured or calculated output after running the optimization algorithm, a new set of optimized yield was determined by putting these final input variables in the multiple

linear regression model. The optimized output calculated as Y_m was obtained by substituting the set of new final values for the input variables into the model equation.

The percentage accuracy (efficiency), η, of the experimental results was calculated using equation 3.

$$\eta = \frac{r_m^2}{r_o^2} \tag{3}$$

where r_m^2 = determination coefficient for the model profile and r_o^2 = determination coefficient for the optimization.

The total methane yield (U_T) derived from the model input variables for the retention period was obtained from the developed equation 4.

$$U_T = \sum_{n=1}^{n=30} [\alpha_0 + \alpha_1(pH)_n + \alpha_2(COD)_n + \alpha_3(NH_4 - N)_n$$
$$+ \alpha_4(T_D)_n + \alpha_5(T_A)_n] \tag{4}$$

where n refers to number of days and each parameter (variable) was measured daily.

Hence, equation 4 reduces to a simplified output equation as shown in equation 5:

$$U_T = \sum_{n=1}^{n=30} (Y_m)_n \tag{5}$$

The total methane yield (W_T) obtained from optimized input variables was obtained using equation 6.

$$W_T = \sum_{n=1}^{n=30} [\alpha_0 + \alpha_1(\overline{pH})_n + \alpha_2(\overline{COD})_n + \alpha_3(\overline{NH_4 - N})_n$$
$$+ \alpha_4(\overline{T}_D)_n + \alpha_5(\overline{T}_A)_n] \tag{6}$$

where:

$\overline{pH}$ = optimized pH

$\overline{COD}$ = optimized chemical oxygen demand

$\overline{NH_4 - N}$ = optimized ammonium nitrogen

$\overline{T}_D$ = optimized digester temperature

$\overline{T}_A$ = optimized total alkalinity

Hence, equation 6 reduces to a simplified output equation as shown in equation 7:

$$W_T = \sum_{n=1}^{n=30} (Y_o)_n \tag{7}$$

where n = retention time in days and Y_o = optimised output.

Table 5 shows the input variables, scaling coefficients and scaling values for the donkey dung. In this model, the

Table 5: Model parameters for donkey dung.

Input variables	Symbols	Scaling coefficients	Scaling values
PH	pH	α_1	−0.267079
Chemical oxygen demand	COD	α_2	0.000004
Ammonium-nitrogen	NH$_4$.N	α_3	−0.000916
Digester temperature	T_D	α_4	0.012151
Total alkalinity	T_A	α_5	−0.000597
Constant		α_0	5.865164

effect of ambient temperatures was considered insignificant because of insulation of the biogas digester.

Substituting the values in Table 4 into equation 2, the optimized model equation for donkey dung is given by:

$$Y_m = 5.865164 - 0.267079(pH) + 0.000004(COD)$$
$$- 0.0009162(NH_4 - N) + 0.012151(T_D)$$
$$- 0.000597(T_A) \tag{8}$$

Figure 12 shows experimental and optimized values for donkey dung. From Figure 12, it is observed that the model did not fit all the experimental data points. Hence, the input values for the model were optimized. The determination coefficient for the model (r_m^2) is 0.9869 and the determination coefficient for the optimization (r_o^2) is 0.9870. From equation 3, this gives a percentage of accuracy of experimental results, $\eta = 99.99\%$.

From Figure 12 it can be observed that the maximum measured methane of 0.3200 m^3 occurred on day 16, and this corresponds to pH = 6.8; COD = 29011; NH$_4$-N = 952; T_D = 32 and T_A = 6442. From the model, the maximum methane yield corresponds to day 16 with the following input parameters:

$$\overline{pH} = 6.7998; \quad \overline{COD} = 29010.19; \quad \overline{NH_4 - N} = 951.99;$$
$$\overline{T}_D = 32.00 \text{ and } \overline{T}_A = 6441.99.$$

This optimized model equation is observed to adequately describe methane production for the biogas digester and has a good fit of determination coefficient, R^2 value of 0.999. The experimental and model values are in agreement and the results express complementary information

to the performance of the batch biogas digester. The pH, NH$_4$-N and T_A have negative values, while COD and T_D have positive values. This means COD and T_D have the greatest impact on methane yield, unlike pH, NH$_4$-N and total alkalinity.

The negative value of pH shows that the methanogens work on a narrow pH range. When pH is increased beyond the range, it has a negative effect on the methane yield; hence, the low methane yield. Similarly, NH$_4$-N has an inhibitory effect (low methane yield) when is greater than 1500 mg/L. The inhibition increases with rising pH value (Deublein and Steinhauser 2008). In an environment, where donkey dung is the only source of digester feedstock, the optimized model equation would be used to obtain an optimum methane yield.

Conclusion

The 1m^3 batch field biogas digester was designed and constructed. During the construction steps, it was ensured that the desired mixing ratio of cement to water for mortar was followed. The plastering layers both inside and outside the digesters which were done in stages increased the strength of the digester walls and sealed air spaces on the digester walls. The epoxy paint used was thick and sealed all air spaces that remained after plastering. After fitting the digester dome on the main body of the digester, it was observed that the main source of gas leakages was where the dome sat on the main body of the digester. However, the correct mixture of water to cement for mortar and the correct type of paint overcame the problem. The insulation of the biogas digester with sawdust minimilised fluactuaclions of temperatures in the batch biogas digester. The designed biogas digester was sucessful because it produced biogas with an average methane yield of 55% when fed with donkey dung as a single substrate. The designed and installed cylindrical batch biogas digester is very suitable in rural communities of South Africa because it is easy to design, construct, feed, agitate and clean.

The optimized model equation for donkey dung is given by:

$$Y_m = 5.865164 - 0.267079(pH) + 0.000004(COD)$$
$$- 0.0009162(NH_4 - N) + 0.012151(T_D)$$
$$- 0.000597(T_A)$$

In conclusion, in an environment where donkey dung is the only source of digester feedstock, the optimized model equation would be used to obtain an optimum methane yield. It can be concluded that the developed

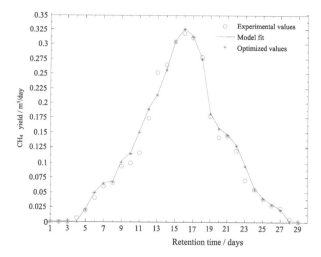

Figure 12: Experimental and optimized values for donkey dung.

mathematical model is applicable for other cylindrical batch biogas digesters provided that the important construction features such as type of insulation, dome structure and wall thicknesses are kept the same.

It can also be concluded that the construction of such biogas digesters in rural communities of South Africa would create employment in local communities, and provide fuel for cooking, lighting and heating purposes. Less carbon dioxide emissions would be released into the atmosphere since people would use biogas and less firewood for cooking.

Recommendations

This research successfully designed and constructed a biogas digester and developed a data acquisition system that was used to measure the performance of the batch biogas digester when fed with donkey dung. A mathematical model was developed to show the behaviour of the biogas digester. It has been noted that more research work should be carried out to establish the relationship between, biogas, slurry and ambient temperatures when a digester is insulated and not insulated. Furthermore, more research work needs to be carried out to determine the bacteria genera that resist temperature fluctuations in a biogas digester and still produce biogas. In addition, further research should be done on co-digestion using field batch biogas digesters to come up with more reliable information. Lastly, the economic aspects of the anaerobic digestion process using different kinds of substrates should also be considered in the future.

Acknowledgements

The authors acknowledge the National Research Foundation (NRF) and University of Fort Hare (UFH) for the support to carry out the research work.

Disclosure Statement

No potential conflict of interest was reported by the authors.

References

ALPHA (American Public Health Association). 2005. *Standard Methods for Examination of Water and Waste Water*. 21st ed. Washington DC: ALPHA.

Cheng, J. 2009. *Biomass to Renewable Energy Process*, 151–163. New York: CKC Press.

Deublein, D., and A. Steinhauser. 2008. *Biogas From Waste and Renewable Resources. An Introduction*. Weinheim, Germany: WILEY-VCH Verlag GmbH & Co.

Dobre, P., F. Nicolae, and F. Matei. 2014. "Main Factors Affecting Biogas Production - an Overview." *Romanian Biotechnological Letters* 19 (3): 9283–9296.

DOE. 2015. Department of Energy. "State of Renewable Energy in South Africa." Matimba House, Pretoria: DOE.

Etuwe, C. N., Y. O. L. Momoh, and E. T. Iyagba. 2016. "Development of Mathematical Models and Application of the Modified Gompertz Model for Designing for Batch Biogas Digester Reactors." *Waste and Biomass Valorization* 7 (3): 543–550.

Lema, J. M., and F. Omil. 2001. "Anaerobic Treatment: a key Technology for a Sustainable Management of Wastes in Europe." *Water Science and Technology* 44 (8): 133–140.

Lettinga, G. 2001. "Digestion and Degradation, air for Life." *Water Science and Technology* 44 (8): 157–176.

Lou, X. F., and J. Nair. 2009. "The Impact of Landfilling and Composting on Greenhouse gas Emissions – A Review." *Bioresource Technology* 100: 3792–3798.

McCarty, P. L. 2001. "The Development of Anaerobic Treatment and its Future." *Water Science and Technology* 44 (8): 149–156.

Mukumba, P., G. Makaka, and S. Mamphweli. 2016a. "Anaerobic Digestion of Donkey Dung for Biogas Production." *South African Journal of Science* 112 (7/8): 1–4.

Mukumba, P., G. Makaka, and S. Mamphweli. 2016b. "Biogas Technology in South Africa, Problems, Challenges and Solutions." *International Journal of Sustainable Energy and Environmental Research* 5 (4): 58–69.

Mukumba, P., G. Makaka, and S. Mamphweli. 2017. "Biogasification of horse dung using a cylindrical surface batch bio-digester." *Intech Open Access–Frontiers in Bioenergy and Bio-fuels* Chapter 21: 425–441.

Mukumba, P., G. Makaka, and S. Mamphweli. 2018. "Mathematical Modelling of the Performance of a Biogas Digester Fed with Substrates at Different Mixing Ratios." *Asian Journal of Scientific Research* 11: 256–266.

Naidoo, R. 2011. *Growth and Development Strategy 2040, Joburg my city, future*.

Nijaguna, B. T. 2002. *Biogas Technology*. New Delhi: New Age International (P) Limited, Publishers.

Rajendran, K., S. Aslanzadeh, and M. J. Taherzadeh. 2012. "Household Biogas Digesters – A Review." *Energies* 5: 2911–2942.

REEIN. 2012. "Practices in Bangladesh: Design of biogas plant." Bangladesh: Renewable Energy & Environmental Information Network.

Santerre, M. T., and K. R. Smith. 1982. "Measures of Appropriateness: The Resource Requirements of Anaerobic Digestion (Biogas) Systems." *World Development* 10: 239–261.

Sharma, N., and G. Pellizzi. 1991. "Anaerobic Biotechnology and Developing Countries—I. Technical Status." *Energy Conversion and Management* 32: 447–469.

Tiepelt, M. 2013. "Business opportunities and barriers in waste in South Africa." SABIA.

Vintilă, T., M. Dragomirescu, V. Croitoriu, C. Vintila, H. Barbu, and C. Sand. 2010. "Saccharification of Lignocellulose - with Reference to Miscanthus - Using Different Cellulases." *Romanian Biotechnological Letters* 15 (4): 5498–5504.

Walker, L., W. Charles, and R. Cord-Ruwisch. 2009. "Comparison of Static, in-Vessel Composting of MSW with Thermophilic Anaerobic Digestion and Combinations of the two Processes." *Bioresource Technology* 100 (16): 3799–3807.

Conclusion: Mathematical modelling and engineering design to promote STEM education in Africa

Mammo Muchie and Nnamdi Nwulu

Mathematical modelling and engineering design need to be part and parcel of the science, technology, engineering and mathematics (STEM) education delivered to all citizens without any exclusion from kindergarten to tertiary level, including all the people in Africa.

There is a need to recognize that mathematical modelling has to include the integration and transformation of science, technology, engineering and mathematics into transdisciplinary, academically rigorous and adjunct forms of innovative knowledge production rather than remaining in the silo of only those within the field of mathematics. Nothing else will bring sustainable solutions to the complex problems Africa continues to face. A new mindset and paradigm shift is needed to go beyond mathematical modelling applied only to the mathematical rendition of different cases by the inclusion of a variety of rigorously articulated and well-defined STEM variables. Mathematical modelling and engineering design have to be re-learned and re-thought to appreciate the current recognition that STEM is now evolving as a result of the integration of science, technology, engineering, and mathematics into a consilience and new unity of knowledge where crossing borders amongst disciplines as a resource for learning and invention is seen as very relevant and significant. The application of mathematical modelling to undertake action, problem, project and work-integrated learning and knowledge production is much needed.

STEM education as a transdisciplinary field is new, but integrated knowledge resources can be enriching for learners to make deeper sense of the world rather than observing through a narrow disciplinary lens bits and pieces of phenomena. Science is defined as a rigorous exploration to discover, understand and know humans, space, land, air, water, nature and all dimensions of reality in the universe. Technology is defined as the techniques developed to design and modify the natural world to facilitate and achieve health, social, economic, knowledge, human vision, and mission objectives. This will enable all people to live decent and wellbeing-anchored lives, managing risks with earlier anticipation and making forecasts for the prevention of any disaster. Technology has become explosive with the current digital cyber world moving ahead of the physical world. It is with creativity, invention and innovation that technology can be used to cure diseases and improve agro-processing production through the creation of a variety of healthy foods and fibers.

Technology has facilitated global communication via the digital internet of things, artificial intelligence, quantum computing, 3D printing and other technology forms. The speed of global communication is increasing with faster data speeds and varied sources of solar, wind and other types of energy. New products, new structures, better housing and better futures through technological applications are now increasing. Technology can be used to reduce the risk of climate change and create the vaccines needed to deal with the current COVID-19 pandemic.

Engineering is a practical application that yields economic, social, health, environmental and novel products through using the knowledge of the mathematical and natural sciences generated through systematic and rigorous research and study. Through consistent application of civil, industrial, electrical and nano-advanced materials, engineering sectors help to bring about sustainable benefits and security for humans and nature. Engineering knowledge and natural materials enable humans to work with tools rather than having to do the hard work themselves, or through the use of animals and plants.

Mathematics is defined by the *Webster's Ninth New Collegiate Dictionary* as any patterns and relationships represented as the science of numbers and their operations, interrelations, combinations, generalizations and abstractions, and the structure, measurement, transformations and generalizations of space configurations. Their specific definitions demonstrate that science, technology, engineering and mathematics are all interlinked as STEM. Together they reinforce each other with the integration of science used to understand how to explore and understand the universe; technology used to facilitate what science and mathematics can produce and bring about change; and engineering enabling modifications to the existing reality by being the defining primary designer and modifier in all the Industrial Revolutions such as the first, second, third and fourth.

The fifth Industrial Revolution is on the way. The application of mathematics and science within all engineering fields uses mathematics that models reality with conceptual frames, variables, numbers, equations and geometry by creating the necessary patterns and relationships. They are combined as STEM to help humans to learn and obtain a deeper understanding, enabling a modification of reality to discover evidence-anchored ways to enrich the knowledge universe rather than each discipline moving along isolated trajectories that may not define, explain and acknowledge fully the real world as best as it can be.

Finally, mathematical modelling and engineering design need to be applied to embed a total invention, innovation and learning culture, which involves transforming Africa through knowledge-competence-building systems using STEM. STEM can be used to combine the knowledge of the community with knowledge from the academy and also with the public and private sectors. Mathematical modelling, engineering design and STEM can become the African language for invention, innovation, learning and capability building by making the priority the transformation of the African education landscape. This will involve inclusion of the very excluded and still unacknowledged and unrecognised past African STEM knowledge to promote African smart, integrated and sustainable development in the fourth Industrial Revolution.

References

Gillispie, Charles C. 1998. "E. O. Wilson's Consilience: A Noble, Unifying Vision, Grandly Expressed." *American Scientist* 86(3):280–283.

Hazelrigg, G. A. 1999. "On the Role and Use of Mathematical Models in Engineering Design." *Journal of Mechanical Design* 121(3):336–341.

Patra, Swapan, Mammo Muchie. 2017." Engineering Research Profile of Countries in the African Union." *African Journal of Science, Technology, Innovation and Development* 9(4):449–465.

Van de Walle, John A. 2016. *Elementary School Mathematics: Teaching Developmentally*. Pearson Education Limited.

Wilson, Edward O. 1998. *Consilience: The Unity of Knowledge*. Vintage Books.

Index

Note: Page numbers in **bold** and *italics* denote tables and figures, respectively.